Analyzing Green Belt Movement's Impact

NICK M. BOWNE

<h1 style="text-align:center">ACKNOWLEDGMENTS</h1>

This dissertation would not have been possible without the support, encouragement, belief, and humor of

… my loving husband, Phillip. Words cannot begin to express how blessed I am for calling you my partner. I would not have been able to even conceive of the idea of getting a Ph.D. without you. You never doubted me throughout this entire process. Whenever I questioned my choice, you reminded me why I chose this path. Whenever I wanted to quit, you pushed back. Thank you for putting up with my moods, my schedule, and 10 years of school! You are my best friend, the love of my life, the person I want to wake up next to in the morning, and the person I want to grow old with.

… my family. Ihr seid immer für mich da und habt immer an mich geglaubt. Selbst wenn ich unleidlich bin oder unzugänglich für eure Vorschläge, oder wenn ich meine Sorgen vor eure stelle, ihr haltet immer zu mir. Die Entfernung zwischen uns macht es mir manchmal schwer die Schwester, Tochter und Enkelin zu sein, die ich gerne wäre. Ohne eure Unterstützung, eure Zusprache und eure Liebe hatte ich es nie so weit geschafft. Danke, dass ihr an mich glaubt und dass ihr meine Entscheidung zu Phillip zu ziehen unterstützt habt; dass ihr mich ermutigt habt, gegen den Strom zu schwimmen und für das zu khämpfen, an das ich glaube; und dass ihr mich an die Realitäten erinnert, die sich ausserhalb meiner kleinen Welt befinden. Dank euch, bin ich ich.

… my mentors. I had the great fortune of meeting two of the most amazing mentors one can imagine during my undergraduate degree at Eastern Michigan University: Brendan Kelly and Ray Quiel. Brendan, thank you for luring me into Forensics. Without you I would have probably never discovered the joy of public speaking, nor would I have developed the ambition to follow my dreams to teach at the university level. Seeing you succeed and following your dreams is the best inspiration I could have asked for (and making it through life without blisters on my hands from rowing certainly has its benefits). Ray, thank you for challenging me, for not allowing me to content myself with faulty arguments, for introducing me to rhetorical criticism and the joys of exploring rhetoric. I hope that when I grow up, I grow up to be as inspirational and supportive to my students as you are to yours. EMU has become a home away from home in large portion because of you.

... my friends, Anglesia, Nick, Steve, Richard, Amy, Lauren, Katharina, Marylou, Vanessa, Kristen, Bill, Jörg, Nadine. Thank you for not giving up on me. I know that I have not been the best friend over the last three years. I have not been in touch much with some of you and have leaned too much on others. Thank you for being the voice of reason, talking me down when I got on my soapboxes, making me laugh when I wanted to cry, being my sounding board when I got stuck; simply said, thank you for being my friends.

... my committee members. Kelly and Ron, thank you for taking me on as your advisee. I know that you both have many responsibilities. Advising someone who has set her mind to finish the dissertation in her third year certainly has made this process stressful for you as well. I truly appreciate that you took a chance on me and continued to encourage me when I questioned myself. I also would like to thank you for agreeing to co-advise me. Receiving feedback from both of you has been very helpful. Sandy, your classes have broadened my horizon more than most I have taken during my academic career. I thank you for introducing me to critical pedagogy and your support in using it as a methodological tool. Dr. White, without your social movement class, I likely would not have chosen the Green Belt Movement as my dissertation topic. While I have had continued interest in the GBM since Maathai's Nobel Peace Prize, I never even considered a social movement study for the dissertation. I am glad I did.

… last, but certainly not least, my four-legged, tail-wagging, barking study buddies, Riley and Griffy. Having our dogs around during the writing process probably saved my sanity. Whenever I wanted to take the computer and toss it out of the window, Griffy brought me a toy to toss instead. When I felt like curling up into a ball and crying, Riley stood above me, staring at me, wagging her tail and kissing/licking me to death. They forced me to leave the house, get fresh air, and remember to take care of somebody else. Their unadulterated joy of seeing us brings a smile to my face every time.

TABLE OF CONTENTS

CHAPTER 1 - INTRODUCTION

There was a horrific forest fire. All the animals ran out of the woods to safety. But the hummingbird filled its tiny beak with water and flew back into the smoke, dropping its mouthful over the flames and returning for another load. The other animals laughed at this minuscule effort. "What do you think you are doing?" the animals mocked. Poised for yet another trip over the fire, the hummingbird replied, "I'm doing what I can." (Wangari Maathai, as cited in Little, 2006)

When the Nobel Committee announced on October 8, 2004, in Oslo that it had decided to award its Peace Prize to an environmentalist, many people were perplexed. Unable to fathom what an environmentalist and a peace prize had in common, these observers were surprised to learn that an environmentalist won during a time when the war against terrorism was raging in Iraq and Afghanistan, and North Korea and Iran were attempting to procure nuclear weapons arsenals. Besides this unusual international context, there was a record field of 194 nominees, among them U.S. President George W. Bush, British Premier Tony Blair, and Pope John Paul II ("Women and Gender," 2004). Despite these impossible odds, the 2004 Nobel Peace Prize went to Wangari Maathai from the Green Belt Movement (GBM) in Kenya. In its press release, the committee noted that the prize was awarded to Maathai

> for her contribution to sustainable development, democracy and peace. Peace on Earth depends on our ability to secure our living environment, Maathai stands at the front of the fight to promote ecologically viable social, economic and cultural development in Kenya and Africa. She has taken a holistic approach to sustainable development that embraces democracy, human rights and women's rights. She thinks globally and acts locally. ("Press release," 2004, October 8)

Though the committee gave the prize to Maathai in particular, it recognized the achievements of the GBM she founded and has headed for over 30 years. While easily mistaken for "just" an environmentalist organization, the GBM is also engaged in promoting women's rights, human rights, education, sustainable development, democratic participation, and peace, among other issues. The movement has taken this broad approach because it views oppression, inequality, environmental destruction, and political instability as interconnected problems. As Maathai put it, "[O]nce you start making these linkages, you can no longer do just tree-planting. When you start working with the environment, the whole arena comes: human rights, women's right, environment rights, children's rights...everybody's rights" (French, 1992). Because these problems are intertwined, the GBM seeks to address them with active educational practices that aim to achieve broad social change. Obi (2005) posited that, particularly on the African continent, the combination of environmental, political, and feminist appeals has been a vital factor in the GBM's efforts to stimulate international pressure on African governments. While the Green Belt Movement is not the only African organization relying on this type of approach, it is one of the more prominent ones. It has achieved global recognition with Maathai's Nobel Peace Prize award and its involvement in the United Nations Environment Programme's Billion Tree Campaign. Furthermore, unlike intranational movements, the Green Belt Movement has been at the forefront of training community leaders from other African nations, using its approach as a model to stimulate tree planting and educational programs. To better understand the GBM's approach to social change, this study investigates the development of the GBM's rhetorical frame and the interaction of that frame with the GBM's educational praxis. The present chapter functions as the introduction to the study by briefly describing the history of the GBM, justifying the study, providing a theoretical foundation to social movement studies, discussing the research questions, explaining the texts chosen for the analysis, and previewing the layout of the dissertation.

History of the Green Belt Movement

During the early 1970s, Wangari Maathai returned from studying in the United States and Germany to find that the entire Kenyan landscape had changed dramatically. Vast stretches of formerly lush, green land were barren and desert-like. Later she would tell reporters that "in Kikuyu, my mother language, there's no word for desert. Yet today much of our land is parched" (Cook, 2005). She also noticed that the resulting lack of firewood, fruit-bearing trees, and nearby water supplies meant the women could no longer provide for their families. In response, she founded Envirocare Ltd., a company designed to provide women with seedlings so that they could grow indigenous trees to improve their food sources, supplement their incomes, and have readily available firewood. The project failed and was nearly buried when, in 1976, Maathai was approached by one of Kenya's elite women's societies, the National Council of Women in Kenya (NCWK), which helped her found Save the Land Harambee, now known as the Green Belt Movement.

On June 5, 1977, Save the Land Harambee planted its first seven trees to mark the worldwide celebration of World Environment Day and dedicated them to deceased Kenyans who had made outstanding contributions at the community and national levels (Maathai, 2007). This strategy has become standard procedure for tree-planting ceremonies, as has reciting the following commitment declaration:

> Being aware that Kenya is being threatened by the expansion of desert-like conditions; that desertification comes as a result of misuse of the land and by consequent soil erosion by the elements; and that these actions result in drought, malnutrition, famine, and death; we resolve to save our land by averting this same desertification through the planting of trees wherever possible. In pronouncing these words, we each make a personal commitment to save our country from actions and elements which would deprive present and future generations from reaping the bounty of resources which is the birthright and property of all. (Maathai, 2004a, p. 21)

Yet, despite their importance, these symbolic tree-planting ceremonies do not form the movement's centerpiece.

At the heart of the GBM are everyday tree plantings by everyday women fulfilling basic survival needs. Shortly after the GBM was founded, the organization asked the Department of

Forestry to provide it with access to 15 million tree seedlings, one for each Kenyan. Though the Conservator of Forests laughed at the group and its plan, he agreed to provide the project with seedlings from the government-run nurseries at no cost. One year later, he recanted his agreement because his department could no longer keep up with the numbers needed by the GBM (Maathai, 2004a). Since then, the Green Belt Movement has taught women to run their own nurseries and grow their own seedlings, providing them with additional income. Furthermore, the GBM has become involved in civic education seminars, planted over 45 million trees in Kenya alone, sponsored the United Nations Environment Programme's Billion Tree Campaign, and engaged in promoting women's rights, Kenya's democracy campaign, food security, and environmental education for schoolchildren.

Justification of Study

This study is justified for a number of reasons. First, because the GBM is a non-Western social movement, analysis of its rhetorical frames provides unique insights. Discussing the emergence of grassroots efforts in the Third World, Escobar (1992) advanced the notion that only a small number of scholars had expressed interest in social movements beyond the westernized world. He argued that much of this lack of interest was rooted in the rhetoric of development, which gives a distorted image of the definition of advancement in places beyond industrialized countries. He posited that, instead of creating a realistic image, "development has functioned as a mechanism for the production and management of the Third World in the postwar period" (p. 413). This idea is supported by Tilly's (2007, as cited in Press, 2009) suggestion that the study of political opportunity, or social movement, is indeed "Western-saturated." While Tilly is one of sociology's foremost movement scholars, this saturation can also be seen in rhetorical studies. Pinkerton (2009) contended that many social movement studies in the field of communication have focused on U.S. and European movements, ignoring much of the rest of the world. Instead, they explore the rhetoric of well-established Western movements, such as the black power movement, the student movement, and the women's liberation movement, amongst others. Conducting a rhetorical analysis of the 2007 Burmese protests, Pinkerton (2009) discovered that social movement theory, because of its cultural restrictiveness, fails to properly address a variety of factors that played a crucial role in the Burmese protest, such as "non-oppositional rhetorical forms of communication, non-modernist conceptions of time, the possibility of collectivities functioning as distributed networks, and the complexity of identification with the movement" (p. 27). By studying the frame(s) employed by the Green Belt Movement, this project attempts to increase our understanding of non-Western social movements and their approach to social change.

Second, this study explores the GBM's educational praxis. While the communication discipline's interest in social movements is strong and varied (Jensen, 2006), there has been a lack of research on education (Holst, 2002). Discussing the connection between radical adult education and social movements, Holst (2002) asserted, "[A]s anyone involved in a movement knows, knowledge is a tool for the important work of political praxis. In other words, we change people's minds in order that we may use that knowledge to change the world" (p. 83). In movements, this education has to occur on two levels: Members and potential members need to understand their positions in society, and society at large has to realize the importance of social change. Yet, while knowledge construction is crucial to movement success, its importance is often dismissed, both by the movements themselves and by researchers. In fact, Foley (1999) argued that "while systematic education does occur in some social movement sites and actions – it is tacit, embedded in action and is often not recognized as learning" (p. 3). Even though the phenomenon has been known as "education by collision" since the mid-1800s, few movement scholars have paid much attention to the role of education in social movements. This study attempts to address this lack of attention.

Third, the Green Belt Movement is significant not only because it is a non-Western movement, but also – and more specifically – because of its geographical location on the African continent. Press (2009) argued that African social movements have been undervalued and understudied, and that their exploration can provide movement scholars with invaluable insights as well as further social movement theory. Exploring sub-Saharan environmental movements, Obi (2005) explained that interest in African environmental movements emerged only after the collapse of the Soviet Union and the subsequent realization that environmental factors impact global security. More importantly, though, the author claimed that because a predominant school of thought in the Westernized world has painted Africa as overpopulated, disease-ridden, and violence-prone, "scholars and policy makers have found it necessary to bureaucratize and depoliticize the emerging environmental movements so that they do not threaten vital Western economic interests in Africa or challenge in any meaningful manner the negative labeling of Africa in the media and official circles" (p. 1). In that regard, it is crucial that African movements such as the Green Belt Movement be included in scholarly research to shed light on the strategies and tactics employed in the face of such adversity.

Lastly, because the GBM is a comprehensive movement organization, an examination of its broad and interconnected messages and frames may help to inform the ongoing debate between movement scholars about the definition of movement organizations as either old (OSM) or new (NSM). Although developing a new definition or classification for social movements is not the primary goal of this study, this study can contribute to the understanding of the phenomenon of social movements at large as well as the theory guiding its studies.

Foundations of Social Movement Theory

Rhetoricians have long worked to justify the study of social movements in the field of communication by trying to create a distinct theory and method for the study of social movement rhetoric. Despite various theoretical endeavors, this has proven difficult. The main reason is that social movements are too complex and not unified enough to constitute a unique genre of rhetorical criticism. If social movements were a unique genre of rhetorical criticism, then the critic could apply certain standards of analysis to the rhetoric. In articulating a theory of genre, Campbell and Jamieson (1978) posited that "genre is a group of acts unified by a constellation of forms that recurs in each of its members. These forms, in isolation, appear in other discourses. What is distinctive about the acts in a genre is the recurrence of the forms together in constellation" (p. 408). Yet, it is exactly this recurrence of forms together that makes the argument for a genre challenging in terms of social movements. As Stewart, Smith, and Denton (2007) have pointed out, social movements are large in scope in terms of geography, time, and events. Treating them as a genre leads to a narrowing of approaches because each movement then abides by the recurring forms.

Since Griffin's (1952) seminal essay on the rhetorical component of historic movements, several scholars have taken his call to study the rhetorical patterns as a mandate to establish a separate theory of social movement rhetoric. Two of the more prominent efforts demonstrate that there is no unique genre and show the benefits of a varied approach.

One of the first scholars to take up Griffin's challenge was Cathcart (1972), who furthered Griffin's belief in a rhetorical component by suggesting that in order to conduct useful and effective analyses of social movements, rhetoricians must begin by abandoning historical and socio-psychological definitions. In fact, he argued, a rhetorical definition is imperative not only "as an aid to rhetorical studies, but because movements are essentially rhetorical in nature" (p 86). The same author (1980) also advanced the argument that true social movements are established through the use of confrontational rhetoric – i.e., rhetoric questioning the basic values and societal norms – as well as an exchange of rhetoric between the agitator and the establishment, or what Cathcart (1980) called reciprocity or "dialectical enjoinment in the moral arena" (p. 272). While not all scholars have subscribed to Cathcart's theory, it helped to create the illusion that social movement rhetoric was distinct.

The challenge to Cathcart's suggestion, however, is twofold: Dialectical enjoinment is not unique to social movements, and social movement rhetoric is not limited to the interaction of rhetoric between the agitator and the establishment. If we consider *genre* to mean that the recurrence of a constellation of elements is unique to that entity, then Cathcart's theory falls short. Studying Johnson's War on Poverty, Zarefsky (1980) found that dialectical enjoinment is possible between any two opposing rhetorical forces, whether movements and establishment, or the President and Congress. Furthermore, Cathcart (1980) believed that "only to the degree that a movement is able to continue to confront the system on moral grounds and engender a counter-rhetoric in kind will it remain a movement" (pp. 271-272). This approach overlooks, for example, rhetoric used to mobilize participants and alter the self-perception of members. It also does not address interaction between the social movements and the larger society. Ultimately, movements need to persuade the public to adopt their goals so that the public will begin to support the movements in pressuring the establishment.

The second scholar attempting to outline specific requirements, proposals, and strategies for social movements was Simons (1970), who suggested that researchers should "examine rhetorical processes from the perspective of the leader of a movement: the requirements he must fulfill, the problems he must face, the strategies he may adopt to meet those requirement" (p. 11). This approach raises several caveats. First, it assumes that all movements have a clear organizational structure with leaders and followers. While this may be true for some of the more traditional movements, new social movement scholars have argued that the assumption no longer holds up (see, for example, Whalen & Hauser, 1995). It is actually quite challenging, for example, to determine who are the true leaders of the Tea Party: While certain prominent figures are relatively outspoken in favor of the Tea Party, none of them has yet claimed leadership. Furthermore, focusing all of the attention on the leader again ignores rhetoric produced by the rest of the movement. Simons (1970) claimed that the requirements mentioned could be addressed only by the leader. Yet, this view does not take into consideration what happens in social movement organizations when the movement leader is not present. For example, under "requirement one," Simons suggested that "they must attract, maintain, and mold workers (i.e., followers) into an efficiently organized unit" (p. 3). Turning to the Green Belt Movement, we can see how this requirement is problematic. While Wangari Maathai has been the single leader of the GBM since its inception in 1976, she cannot possibly be at every tree-planting ceremony, nor can she lead each and every nursery. In fact, although the GBM has a centralized structure, mobilization of participants and maintenance of individual villages' efforts are left to the villages (Maathai, 2007). This in turn requires rhetoric to "attract, maintain, and mold workers" by people other than Maathai. If we were to use Simons's (1970) theory to argue for a unique genre of rhetorical criticism, we would have to focus our efforts only on the leader, and would miss relevant and necessary efforts by the rest of the movement.

While there certainly was never complete agreement within the field of communication on what constitutes a social movement, how it functions, and how to study it, the matter was further complicated by the "cultural turn" taken in the early 1980s (Whalen & Hauser, 1995). Since then, scholars have spent much time debating the merits of both schools of thought rather than focusing on the rhetorical significance of individual movement efforts.

Overall, the debate can be read as a change from universal demands (e.g., class and economic struggle) to particularized demands (e.g., identity construction and rights). More traditional social movement scholars have assumed that movements concentrate on achieving improvement for specific material issues, and that social struggle arises out of economic and materialistic inequalities (Cho, 2008). Based on Marxist theories, this branch of social movement scholarship has studied how social movement rhetoric functions to mobilize members in their effort to overcome economic challenges and improve their lives. New social movement scholars, on the other hand, have suggested that a simple improvement in class standing does not address underlying, deep-rooted issues of power imbalance. Therefore, instead of "class," topics of study should include "identity, race and ethnicity, the body, ethics, narrative, technology, textuality, representation, gender, globalization, hegemony, resistance, performance, space and place," amongst others (Carleone & Taylor, 1998, p. 338). Thus, understanding the construction of reality and identity-based topics provides insight into the impact language has on people's perception of themselves in society as well as the creation of power.

Within the Green Belt Movement, elements of both schools of thought are represented. Maathai's primary reason for creating the GBM was to provide women with ways to care for their families (Maathai, 2007). Because of deforestation, the land was parched, and women lacked the resources to improve or at least maintain their lifestyle. Certainly, fighting for food, water, and survival is a very obvious economic and materialistic issue. Yet, at the same time, addressing this universal concern automatically led the GBM to include identity construction. By teaching women how to provide for themselves, it also taught them to rethink themselves in terms of the society at large: to develop a greater understanding of the place, of the politics around them, of the patriarchal structures constricting them (Maathai, 2004b). As such, studying how the GBM frames social change rhetorically provides the communication discipline with the opportunity to broaden its approach to the study of social movements, moving beyond the distinction between old and new social movements.

Research Questions

This study attempts to place the Green Belt Movement within the current definition of social movement theory before examining the frame(s) employed by the GBM as well as its educational practices. The GBM's broad approach to social change was implemented in response to the pressing environmental, women's rights, and economic issues in Kenya. The intent of the study is to further understanding about the impact and effectiveness of such a comprehensive approach for this social movement in particular, as well as its significance for social movement studies in general. In order to achieve this purpose, the study asks the following research questions:

RQ1: How did the Green Belt Movement's frame(s) develop over the course of its existence? The goal of this question is to investigate the development of the frame(s) employed by the Green Belt Movement. As Kuypers (2009) explains comparing and contrasting the frame(s) established by a rhetor in several texts over the course of a designated period of time gives insight into the development of those frame(s). Conducting a textual analysis of major award acceptance speeches, this study outlines the establishment of frames in each instance. The results of each section are then used to delineate the development of the frame(s) over the course of 20 years. The results gleaned from this analysis can illuminate how Maathai framed the GBM approach to social change in an effort to increase the acceptability of that approach.

RQ2: Do the GBM's practices follow critical pedagogical principles? Relying on a textual analysis of two manuals published by the GBM, this question investigates the practices of the GBM using critical pedagogical tenets. As previously suggested, the GBM's objectives range from environmentalism to human rights to democracy efforts. If empowerment and the reduction of oppression are truly at the heart of the GBM's efforts, then this question should reveal the movement's use of critical pedagogical principles.

RQ3: How do the Green Belt Movement's educational practices reflect the frame(s) established? Using the results of Chapter 3 and 4, this question explores whether and how the Green Belt Movement's approach to education accords with its approach to social change as put forth in the frame(s) established.

RQ4: What insights can be gained from the exploration of the GBM's frame and educational efforts with regard to their applicability beyond Kenya? Because of the often narrow approach taken by movement organizations, social movement scholars have grappled with the applicability of social action frames beyond the specific movement under investigation (Benford & Snow, 2000). The Green Belt Movement's unique approach to social change may decrease these questions and increase potential frame transferability. Additionally, this question also addresses how the influence of critical pedagogical principles as strategies for education can advance a movement's cause. Although education and knowledge construction are vital elements for a movement's survival, they are often not overtly articulated as functioning strategies. The GBM's focus on training and education suggests that being more overt about the importance of education could be beneficial to other social movements as well.

RQ5: What are the consequences of the GBM's framing and educational practices for social movement theory? This question attempts to address the challenge of placing a movement organization within the two dominant paradigms available to researchers. By exploring the GBM's approach, this study furthers the argument that the Green Belt Movement does not fit neatly into just one category. Consequently, this study illuminates the limitations current definitions and classifications of social movements theory create for the study of social movement organization. Thus, this study may be able to bridge the gap between traditional and new social movement theory, and may provide a basis for collaboration as well as a useful approach to the study of contemporary social movements.

Texts

To answer these research questions, this study relied on two sets of texts, each of which provides insight into certain elements of the GBM. The first set of texts was used to help answer RQ1 and provide insight into the development of the GBM's frame(s). For this purpose, the study utilized three award acceptance speeches given by the movement's leader, Wangari Maathai. The speeches, delivered about a decade apart, spanned roughly 20 years of the movement's activity. These three speeches were chosen among the many given by Maathai because of their pivotal nature. The first was given in 1984 on the occasion of one of the first awards earned by the Green Belt Movement; the second was presented in 1993 during the Science Festival in Edinburgh, Scotland, at a time when Maathai and the GBM were under heavy attack by the Kenyan government; and the final oration chosen was Maathai's acceptance speech for the Nobel Peace Prize in 2004. Although these documents single out Maathai's rhetorical efforts, they also provide valuable insight into the GBM's framing of issues. As the organization's founder and leader for over thirty years, Maathai has been the source of most, if not all, of its public rhetoric. Maathai has served as the movement's leader since its inception in 1977 and is inextricably connected to its activities. In fact, Ndegwa (1992) has suggested that it is sometimes difficult to separate the GBM and Maathai because of her complete immersion in the movement. Furthermore, as the face of the movement, Maathai has shaped most of the movement's discourse since its inception, including the GBM's perception by international audiences. Thus, her award acceptance speeches lend valuable insights into the development of the GBM's frame(s).

The second set of texts was used to explore the GBM's educational practices and help answer RQ2. Both of the documents are manuals Maathai published to describe the GBM's approach to outsiders. The first manual, *The Green Belt Movement,* was published in 1985; its specific objective was "to document some facts about the movement so that the experience may be more broadly shared with those interested" (preface). This document served as the foundation to explore the GBM's setup and educational approach because it discussed the creation of the movement as well specific facets of its procedures and objectives. The manual was chosen for this study because it is closest available thing to a founding document of the GBM. Although Ndegwa (1992) referred to the Green Belt Movement's constitution in his study about nongovernmental organizations (NGOs) in Kenya, this founding document cannot be publicly accessed. It stands to reason that many of the principles discussed in the constitution were incorporated in this manual.

The second document used in this portion of the study was Maathai's 2004 book *The Green Belt Movement: Sharing the approach and the experience.* This book is an expanded edition of the 1985 manual in terms of time span and content. Because this book covers a longer period of time, it stands to reason that it provides deeper insight into the development of the GBM's educational approach. Additionally, this second edition includes more detailed descriptions of the early efforts of the GBM, and thus serves to supplement or contrast any findings uncovered in the first section of the analysis.

Again, both documents were published by Maathai. Viewing the GBM solely through her perspective could certainly be problematic; but, as already indicated, one of the consequence of Maathai's total immersion in the GBM has been the fact that most of the movement's rhetoric stems from her (Ndegwa, 1992). As the public face of the GBM, she has consistently shaped the international public's image of the GBM through her rhetoric. It is this consistency that makes these two texts appropriate choices to answer RQ2, especially in light of the fact that both texts were published so that the GBM's approach could be implemented across the globe (Maathai, 1985, 2004a).

Organization of Study

The dissertation is divided into five chapters. The first chapter justifies the topic of study, discusses the research questions, and briefly describes the methodological approaches as well as the texts selected for analysis. Chapter 2 explains the methodological tools utilized for the study by reviewing the pertinent literature on framing theory as well as critical pedagogy. Chapter 3 investigates the development of the Green Belt Movement's frame(s), seeking to answer RQ1. Through a textual analysis of the three award acceptance speeches mentioned above, this chapter explores the GBM's frame development. More specifically, after a short background is provided, each speech is analyzed individually. Once all of the individual analyses are concluded, the chapter closes by tracing the development of the frame(s) over the course of the chosen time span. Chapter 4 provides an answer to RQ2 by examining the GBM's education practices for critical pedagogical principles using a textual analysis of the Green Belt Movement's manuals. Again, each text is investigated separately before broader conclusions about the GBM's education approach can be derived. Lastly, Chapter 5 answers all five research questions and offers some insights into their meaning for rhetorical social movement scholarship. This chapter concludes by discussing limitations to the study as well as proposing future research.

CHAPTER 2 - METHODOLOGY

This chapter briefly discusses how rhetorical criticism functions before investigating the placement of social movement studies within the confines of rhetorical analysis. In addition, this chapter explains the basic principles behind framing, as well as its use in rhetorical analysis, before providing insights into the procedures used in this study. Lastly, this chapter describes the theoretical foundation of critical pedagogy, its utility as a tool of analysis for social movement scholars, and the basic tenets used to explore the educational practices of the GBM in this analysis.

Rhetorical Criticism

While traditionally focused on oratory, rhetorical or textual analysis is no longer used exclusively for analyzing speeches. Sillars and Gronbeck (2001) suggested that this method is useful for the study of any text concerned with the evaluation of the best use of persuasive means. "Text" does not refer to a single written speech, but to "anything that influences the values, beliefs, attitudes, and behaviors of the public" (Hunt, 2003, p. 378). In the case of social movement study, this includes, but is not limited to, speeches, leaflets, emails, rallies, manifestos, and any number of symbolic protest acts, such as spraying fur coats with red paint (as done by PETA) or shaming police officers by walking up to them naked (as done by Maathai and other protesting women). By concentrating on the effectiveness of the persuasive means utilized, rhetorical criticism can increase awareness of potential rhetorical choices and maneuvers, "so that others might either incorporate them into their own practice or shy away from them" (Zarefsky, 2006, p. 386). While Zarefsky did not refer specifically to the study of social movement rhetoric in this essay, he did suggest that "rhetorical criticism can be applied to *anything*, so long as the object of the criticism is explained by reference to rhetorical concepts and issues" (p. 386). This notion was also supported by Campbell (2006), who proposed that rhetoric is ubiquitous and indigenous, linked to culture, tradition, and language. She elaborated on this idea by defining "rhetoric" as "the study of language and how language shapes perception, recognition, interpretation and response" (p. 360).

Benefits of Studying Social Movement Rhetoric

The communication discipline's major contribution to social movement scholarship is to further understanding about the impact and effectiveness of the rhetoric used by social movements. Most rhetorical scholars (for example, Cathcart, 1972, 1980; Simons, 1991; Stewart, 1980; Zarefsky, 1980) have agreed that social movements function through persuasion. This means that persuasion is one of the dominant tools available to social movements to achieve their goal of social change. As such, examining the rhetoric of social movements is necessary to gain an understanding of how they shape their strategies of persuasion to achieve their goals. The lenses used to conduct rhetorical criticism, however, need not be limited to a single methodological approach. In fact, Campbell (2006) concluded that approaches have to be appropriate to the character of the rhetoric as well as the cultural context. In proposing a functional approach to the study of social movement rhetoric, Stewart (1980) contended that while the functions are not unique to social movement campaigns, they still further illuminate how the rhetoric helps to propel or retard the movement. Relying on the multiplicity of lenses available to them, rhetorical scholars can explore the impact of the movement's rhetoric. The rhetorical study of social movements, then, provides a catalog of options that address anything from how the movements interact with the establishment (e.g., Bowers, Ochs, Jensen, & Schulz, 2010; Black, 2003), to how the movement help creates self-identity (e.g., Whalen & Hauser, 1995; West, 2007), to the prescription of courses of action (e.g., Bowers et al., 2010). Framing analysis can and should provide such an approach because it is concerned primarily with the ways language helps people negotiate and interpret events.

Although there are specific benefits that can be achieved through studying social movements rhetorically, it is important to recognize their interdisciplinary nature. In fact, to claim that social movements are purely rhetorical in nature ignores the interest shown by sociologists, political scientists, and historians in the development of the rhetorical strategies. As Lucas (1980) suggested, "[M]any sociologists now appear to agree that the ultimate success of social movements depends upon their ability to challenge persuasively prevailing thoughts, beliefs, attitudes and values" (p. 261). Moreover, social movements display such a complex nature that sociological and rhetorical approaches should be seen as complementary rather than oppositional.

Framing

Framing was utilized as the lens through which a textual analysis of the primary texts was conducted to explore the development of the GBM's frame. Because the Green Belt Movement is involved in a variety of issues, touching on environmentalism, human rights, women's rights, education, democracy, and peace, it is crucial to understand how the GBM frames its attitude toward social change, as well as the various elements of its efforts. It is unusual for a social movement or social movement organization to take such a comprehensive approach to social change. Before being able to explore the specific issues, therefore, it was necessary to investigate how this frame impacts the movement's work and/or effectiveness. Thus, the following section clarifies the basic principles of framing and how it was utilized in the present study.

Based on the theories by Bateson (1972) and Goffman (1974), "framing" refers to the way people organize events in the world around them. Goffman (1974) proposed "that definitions of a situation are built up in accordance with principles of organization which govern events – at least social ones – and our subjective involvement in them" (pp. 10-11). This suggests that, in addition to classifying information in a way that allows people to make sense of the world, the language used to construct the frames influences what we see and how we see it (Sillars & Gronbeck, 2001). As Burke (1966) explained, "[E]ven if any given terminology is a *reflection* of reality, by its very nature as a terminology it must be a *selection* of reality; and to this extent it must also function also as a *deflection* of reality" (p. 46, emphasis in original). That means that by choosing to select one aspect over another, a rhetor has already made a decision about how he or she desires the audience to perceive the issue.

The most common metaphor used to describe framing is that of a window or picture frame (based on Bateson, 1972, p. 188). Depending on the placement, size, or angle of the window (or picture frame), a person will have varying views of the same landscape (or picture). Framing functions in much the same way by including or excluding information or by describing an event in certain terms, "inducing us to filter our perception of the world in particular ways, essentially making some aspects of our multidimensional reality more noticeable than other aspects" (Kuypers, 2005, p. 186). Because it makes complex ideas more manageable, this practice allows us to negotiate and interpret everyday events in an interpersonal setting, in news reporting, or as a social movement advancing an agenda of change.

Although often used subconsciously, framing becomes a rhetorical device when it is utilized to influence people's interpretation of a situation. An analysis of these devices then provides an insightful way to understand the impact of rhetoric. Burke (1954) suggested that "[r]hetoric deals with the possibilities of classification in its *partisan* aspects; it considers the ways in which individuals are at odds with one another, or become identified with groups more or less at odds with one another" (p. 22). It is rhetoric that is needed to create unity out of division (Brock, 1985). By providing a new perspective, framing allows the rhetor or the social movement to establish identification with audience members, giving them the opportunity to align themselves with the rhetor or the movement. A rhetor can create this new perspective by selecting certain elements of the event and making them more salient than others (Entman, 1993).

Framing needs to occur in such a manner as not to collide with the audience's perception of reality. According to Burke (1966), "Any new rationalization must necessarily frame its arguments as far as possible within the scheme of 'properties' enjoying prestige in the rationalization which it would displace" (p. 66). This means that while framers strive to make the audience understand a concept differently, they try not to violate the audience's perception of reality to an extent that would lead to alienation and, ultimately, rejection of the frame. Benford and Snow (2000) described the process as follows:

> Frame articulation involves the connection and alignment of events and experiences so that they hang together in a relatively unified and compelling fashion. What gives the resulting frame its novelty is not so much the originality or newness of its ideational elements, but the manner in which they are spliced together and articulated, such that a new angle of vision, vantage point, and/or interpretation is provided. (p. 623)

Because movement organizations such as the GBM attempt to enact social change, framing provides a useful tool to realign audiences' perceptions so that they fall in line with the movement's concerns and goals. At the same time, framing is a particularly interesting tool to study the rhetoric of social movements because of the importance of persuasive discourse to the movements' effectiveness.

How to Conduct a Rhetorical Framing Analysis

As suggested above, rhetorical criticism is concerned with the study of the effectiveness of the rhetorical means used by a rhetor. Kuypers (2009) argued that, in contrast to social scientists, rhetorical critics do not approach their object of study with preconceived hypotheses, but let sometimes vague research questions guide their examination of the specific text(s). Furthermore, a method should be thought of in terms of a perspective or lens that guides the analysis rather than a rigid construct that has to be followed. Each perspective illuminates a different element of the complex rhetorical act. Framing analysis constitutes one of these lenses.

Drawing on Burke's definition of "terministic screens," Ott and Aoki (2002) explained that "frame analysis looks to see how a situation or event is named/defined" (p. 485). Rhetorical critics do this by conducting a close textual analysis during which they investigate the text for key terms, themes, metaphors, or descriptions of people, ideas, and events (Entman, 1993). These themes represent a subject of discussion that is framed in a particular way. The frames emerge as a consequence of the analysis. Because this methodological approach functions inductively and does not rely on a priori categories, critics do not know beforehand the nature of the frames they will find or the number of frames within the rhetorical act. It is entirely possible, then, that some situations rely on a singular frame to advance their agenda, while others employ multiple complementary or even competing frames. Once the frames have been uncovered, the critic constructs an argument about the rhetorical act's impact and effectiveness, using the text itself as supporting evidence.

Kuypers's (2006) study *Bush's war: Media bias and justification of war in a terrorist age* provides a useful example that can help demonstrate how to conduct this type of study. In his book, the author explored how mainstream news media reported on Bush's rhetoric about the War on Terror. For this purpose, the author conducted a comparative analysis between the content of the president's speeches and the content of the news media's reporting of those speeches. Rather than relying on a priori themes or frames, Kuypers discerned the frames through repeated close-textual readings by looking first at the president's speeches and then at the media's reports. Only after discovering the frames in each collection of texts did Kuypers compare and contrast the results (Kuypers, 2009). This method demonstrated that although the media continuously reported on the President's rhetoric, their framing of the issues changed from echoing the president's perspectives to contradicting them and, ultimately, even undermining him. While Kuypers's inquiry did not constitute a social movement study, the same principles apply.

Procedures

To extrapolate the GBM's frame(s), framing was used as a methodological tool to conduct a close-textual analysis of the three award acceptance speeches given by Maathai over the course of 20 years. The time span of the three speeches allowed them to show the development of the frame(s) used. In order to arrive at understanding the development, though, each speech was first explored individually, utilizing Entman's (1993) conception of framing, namely selection and salience.

To begin, each speech was placed in its historical context to outline the importance of the occasion and the award to Maathai and the GBM. Following the brief description of the background, the content and structure of each speech was described. As previously mentioned, Kuypers (2009) suggested rhetorical scholars using framing analysis engage in it slightly differently than social scientists. Instead of investigating texts for a priori categories, rhetoricians conduct close-textual analysis to let sometimes vague research questions guide their analysis of specific text(s). In that regard, framing is a lens through which the rhetor views the artifact in question. This meant that, although the GBM is engaged in a variety of social concerns, including environmental protection, women's rights, sustainable development, democratic participation, and peace issues, the texts were not read for these issues. Instead, each text was read repeatedly to understand what issues Maathai singled out in each of them. Because rhetorical framing analysis does not rely on a priori categories, the actual specific content of the speech speaks to the topics selected by the rhetor. Entman (1993) argued that by selecting certain topics over others, the rhetor begins to shift the audience's focus. The content of the speeches provided the basis for the following framing analysis.

Taking the content into consideration, the object of the framing analysis was to delineate how Maathai wanted the audience to view the GBM's approach to social change. This notion corresponds to Entman's (1993) second tenet: salience, or making some elements more rhetorically important. This means that rhetors discuss topics in ways that suggest to their audiences how they want them to think about those topics. Of importance here is not only the space allotted for these themes, but also the potential rhetorical effect of the description. As such, the content of the speech functions to support the framing argument. Maathai's language choices were then scrutinized to illuminate how the themes helped establish the overall frame(s) of the speeches. The analysis of the three speeches then served as the foundation to explore how the GBM's frame(s) have developed over the movement's existence.

Contributions to the Rhetorical Study of Social Movements

As mentioned in the introduction to this dissertation, framing analysis can make valuable contributions to the rhetorical study of social movements. The three most promising areas are as follows: 1) Rhetorical framing analysis can show how the movement strives to fulfill certain rhetorical functions. 2) This method helps trace the development of the movement's rhetoric. 3)

Through comparative analysis, insights can be gained into the perception of the movement by others, be they the establishment, the media, or the public.

First, framing analysis can show how movements strive to fulfill basic persuasive functions as well as provide judgment as to their effectiveness. Building on Stewart's (1980) article, Stewart et al. (2007) argued that social movement rhetoric needs to fulfill six basic functions: transforming the perception of social reality; altering the self-perception of protestors; legitimizing the social movement; prescribing courses of action; mobilizing for action; and sustaining the social movement. Each of these functions requires that the movement provide its interpretation of the issue/reality, and their effectiveness depends on the articulation of successful frames.

For example, the first function, transformation of the perception of social reality, can reinterpret the past, the present, or the future. Over the course of its existence, the Green Belt Movement has touched on all three of these, but for the purpose of example I will focus on how the GBM framed the urgency of the present. When the GBM first started, the most crucial issue was to draw attention to the situation of women in Kenya as well as to a possible solution: the fields were eroding *now*, women and children were *now* malnourished, and streams were *now* drying up (Maathai, 2007). Although the obstacles seemed nearly insurmountable, the solution was simple: plant trees. As Maathai (2007) put it, "I responded to the needs of women. Their issues were clean water, firewood, wood to build their shelter, and they needed nutritious food. And everything they were asking of, I connected to the environment... So, I told the women, 'But we can plant trees.'" All of the issues facing the women in Kenya were linked to the felling of indigenous trees. Hence, the destruction of the natural environment was the GBM's frame in addressing this persuasive function.

In addition to showing how frames can help articulate the persuasive functions, this brief example demonstrates that instead of introducing radically new thoughts and ideas, the GBM relied on existing beliefs, attitudes, and values, reflecting Benford and Snow's (2000) claim that a frame's novelty lies in its ability to provide a new angle to existing events and experiences. By addressing how movements strive to fulfill the persuasive functions, critics can explore the construction of the frames and judge their effectiveness based on their acceptance potential.

The second major contribution of framing analysis to rhetorical social movement studies is that it helps trace the development of the movement's rhetoric. This involves a comparative analysis of various pieces of movement rhetoric discussed in this dissertation. Since there are roughly 10 years between each of the three speeches under consideration, comparing their frames demonstrates how the movement's rhetoric has developed and evolved over the years. This type of insight is important for rhetorical critics because it allows them to evaluate how movements adapt to change and time constraints. Considering that time is one of the greatest challenges for most movements (Stewart et al., 2007), scholars and activists alike can benefit from understanding how the frames of successful or unsuccessful movements have contributed to their survival or demise.

One theory that might be of particular interest in this regard is that of Snow, Rochford, Worden, and Benford (1986), who suggested that framing in social movements usually occurs in one of four contexts. *Frame bridging* refers to the linkage of two or more ideologically congruent but unconnected frames. *Frame amplification* highlights or clarifies an organization's principal theme by focusing particularly on its values or beliefs. *Frame extension* provides movements with the opportunity to broaden the primary frame by including points of view that are congruent but not yet officially included. *Frame transformation* allows the movement to reshape old meanings and erroneous beliefs. Although the authors focused particularly on participant mobilization, the four contexts might help to explain in more detail how a movement's frames change over the course of time and how specific circumstances require specific frame adaptations.

Lastly, comparative framing analysis can provide insights into the perception of the movement by others, such as the establishment, the media, and/or the public. Much as Kuypers (2006, 2009) compared a rhetor's primary rhetoric – President Bush's speeches – with the media's reactions, social movement scholars can engage in the same practice. This type of analysis can take a variety of forms to answer a variety of questions. For example, a scholar could analyze the frames constructed by the Tea Party Movement and compare them with the frames reflected by mainstream news media, contrast conservative and liberal news media, or look at the reactions by governmental officials. One could also compare the rhetoric of orators frequently appearing at Tea Party Movement rallies with the reactions of movement members on websites and blogs. While the specific questions guiding each of these studies vary, all of them are oriented toward understanding how the frames constructed by a social movement are reflected and perceived by the broader society. Considering that social movements strive to enact social change, understanding how the framing efforts are received by society is a crucial element to success. Framing analysis can therefore help rhetorical critics to answers questions of perception.

Critical Pedagogy

In order to answer RQ2, a textual analysis of two manuals written by Maathai (1985, 2004a) was conducted. Utilizing tenets derived from Freire (1970), this analysis explored whether and how the GBM's educational efforts reflect critical pedagogical principles. The following section illuminates some key concepts of critical pedagogy and discusses the specific element that was applied to this project.

At its most basic, critical pedagogy is a teaching philosophy intended to empower the marginalized and disenfranchised by focusing on the student's lived experience rather than a teacher-centered and text-based curriculum (Glenn, 2002). By drawing from the student's life, culture, and language, it allows the subject matter to become more meaningful and to provide a frame for examining sociopolitical constructs confining and oppressing the student (Freire, 1970; hooks, 1994; McLaren, 2003). Freire (1970) exposed the traditional educational system as a banking system of education in which teachers deposit knowledge into students without any concern for the information's relevance or usefulness. Instead of advancing knowledge and equality, this educational system perpetuates existing oppressive structures, providing those in power with the means to keep those at the bottom at the bottom.

Based on Freire's work, critical pedagogy is derived from radical and progressive educational movements aspiring to link democratic principles with transformative action (Darder, Baltodano, & Torres, 2009). When oppressive culture and hegemony are demystified through a practical educational approach, students become empowered (McLaren, 2003). Empowerment, for McLaren, meant not "only helping students understand and engage in the world around them, but also enabling them to exercise the kind of courage needed to change the social order where necessary" (p. 85). While McLaren did not explicitly include social movements in his discussion of critical pedagogy, social movements strive to empower their members to the extent that they not only understand but also have the courage to resist and enact social change. As such, critical pedagogy provides a useful tool for exploring the educational practices of social movements in general and the GBM in particular. In terms of this study, understanding the educational practices might give insight into the sustainability of the comprehensive approach to social change the movement employs.

Freire (1970) argued that although teachers are the oppressors in the banking system, they need to become active participants in education as a libratory practice. Hooks (1994) suggested that one of the greatest lessons educators can learn from Freire is how a "critical privileged thinker [should] approach sharing knowledge and resources with those... in need" (p. 53):

> Authentic help means that all who are involved help each other mutually, growing together in the common effort to understand the reality which they seek to transform. Only through such praxis – in which those who help and those who are being helped help each other simultaneously – can the act of helping become free from the distortion in which the helper dominates the helped. (Freire, 1978, p. 8)

Teachers can and should become helpers in the struggle against oppression, but they need to be aware of the danger of simply paying lip service to engaged praxis. Freire (1970) posited that "only by working with the people could [teachers] achieve anything authentic on their behalf" (p. 41). Therefore, it is imperative that teachers leave their classrooms and venture out into the neighborhoods and streets (Scatamburlo-D'Annibale, Suoranta, & McLaren, 2006). This argument also supports the application of critical pedagogy to social movement studies, because many social movements occur outside the classroom and the establishment. The vast majority of discussions of critical pedagogy, however, have indeed been limited to how to employ its tenets in the classroom (e.g., Glenn, 2002; hooks, 1994; McLaren, 1999; Pineau, 2002; Toyasaki, 2007). Cho (2008) even argued that, in addition to giving greater weight to the classroom, "critical pedagogy is relatively weak in embedding the analysis of education in the structure of economy and polity" (p. 4). By bringing critical pedagogy into social movements, activists and scholars alike can learn to employ critical pedagogical practices as strategies for education in an effort to advance their movement's cause.

Despite growing interest in critical pedagogy, there are a few concerns that need to be addressed before discussing the particular tenets used in this analysis. Feminist scholars have critiqued Freire and critical pedagogy as perpetuating sexist language as well as patriarchy and gender roles (Darder et al., 2009). Hooks (1994) explained that much of Freire's sexist language and perpetuation of patriarchy is rooted in his cultural background. At the same time, however, she suggested that "Freire's own model of critical pedagogy invites a critical interrogation of this flaw in the work" (p. 49). Furthermore, despite sexist tendencies, Freire's work is applicable to women's groups, such as the Green Belt Movement, because of his "recognition of the subject position of those most disenfranchised, those who suffer the gravest weight of oppressive forces" (hooks, p. 53), including women.

Ecological critics have argued that critical theory in general and critical pedagogy in particular perpetuate the advancement of Western thought and values by focusing on the notions of humanity, freedom, and empowerment (Darder et al., 2009). When these values are overemphasized, indigenous knowledge and non-Western tradition lose their relevance. Bowers and Apffel-Marglin (2004) suggested that privileging of individual reflection in the name of empowerment stands in stark contrast with those societies valuing community knowledge and collectivity, and may therefore lead to further alienation of human beings from nature. Yet, addressing this critique, Gruenewald (2003) suggested that one solution could be place-based critical pedagogy "that interrogates the intersection between urbanization, racism, classism, sexism, environmentalism, global economies and other political themes" (p. 6). McLaren and Houston (2004) proposed that critical pedagogues must work alongside new social movements to cultivate sites for capacity-building and democratization. Analyzing the GBM's educational practices can help address the ecological criticism by joining place-based critical pedagogy with social movement scholarship.

Procedures

Freire (1970) described three elements needed for critical pedagogy to occur: conscientization, praxis, and dialogue. These three elements were used as the basis for conducting the analysis of the GBM's educational practices. Freire (1970) suggested the first step in achieving education as practice of freedom has to be *conscientization* or, as hooks (1994) translates it, "critical awareness and engagement" (p. 14). This conscientization or awakening is crucial to Freire because the oppressed "identif[y] with the oppressors; they have no consciousness of themselves as persons or as members of an oppressed class" (p. 46). It is the moment when a person starts to think critically about the self and his/her identity in the political surroundings that starts the process of transformation (hooks, 1994). In that moment, the oppressed begins to move from an object to a subject. Investigating the process of conscientization within the GBM can provide insight into strategies of empowerment for social movements functioning in non-Western as well as Western circumstances.

The second element Freire considered significant is *praxis*. Indeed, conscientization is never an end by itself but is inextricably joined with praxis (hooks, 1994): "It is action and reflection" (Freire, as cited in hooks, 1994). This emphasis on praxis suggests that instead of passively depositing meaningless knowledge into the students, the teacher should work to actively engage students, making knowledge meaningful by drawing on the students' lived experience. Drawing on personal experience, critical pedagogy allows for reflection of that experience in relationship to the gained knowledge. Ultimately, the goal of critical pedagogy is to turn the oppressed object into a subject that can take the appropriate action to demand and enact social change.

Both conscientization and praxis can be achieved through the third element, *dialogue*, where students and teacher are jointly responsible for the process of creating knowledge. Freire argued that "knowledge only emerges through invention and re-invention, through the restless, impatient, continuing, hopeful inquiry humans pursue in the world, with the world and with each other" (p. 72). As such, dialogue can be a liberating practice only if the oppressed are actively involved in reflective participation.

While critical pedagogy is more often considered a practical approach to education and theory rather than a methodological tool of analysis, Giroux (1985) illustrated the relevance of textual analysis for critical pedagogy. The author suggested that through analysis of educational practices, the implementation of a critical pedagogy can be advanced. Thus, applying critical pedagogical tenets as a methodological tool to the GBM's manuals provides insight in the organization's educational practices and allows conclusions about their importance to the GBM's approach. Considering the implicit importance of education for social movements and the lack of research in that arena, this study furthers scholars' understanding of education as a core element to social movement praxis.

CHAPTER 3 – FRAMING ANALYSIS

Maathai's Framing of the Green Belt Movement

This chapter delineates the frames used by Maathai in talking about the Green Belt Movement over a span of 20 years. As discussed previously, for that purpose, three award acceptance speeches were chosen, each occurring roughly 10 years apart. In each section of this chapter, a different speech is analyzed, beginning with Maathai's Right Livelihood Awards acceptance speech of December 9, 1984. This is followed by an analysis of the acceptance speech for the Edinburgh Medal in 1993 at the Edinburgh International Science Festival. The last speech analyzed is Maathai's Nobel Peace Prize acceptance speech of December 10, 2004.

To begin, each speech is placed in its historical context to outline the importance of the occasion and the award to Maathai and the GBM. Next, the content and structure of each speech are described. Because rhetorical framing analysis does not rely on a priori categories, the specific content of the speech speaks to the topics selected by the rhetor. As Entman (1993) suggested, selectivity is one of the crucial elements in framing as the rhetor begins to shift the audience's focus by choosing some elements over others. Describing the actual topics of discussion in each speech provides the basis for extrapolating the frame(s) advanced. Finally, each section concludes with the framing analysis. As previously mentioned, the themes of the rhetorical act suggest the frame used by the rhetor. As such, it is crucial to identify how the themes help establish the frame(s) of each speech. In terms of Entman's (1993) primary characteristics of framing, this part of the analysis deals with the element of *salience* – the most important and noticeable element(s) described. By making some element rhetorically more important, the rhetor suggests to the audience how he or she wants them to think about the topic. Of importance here is not only the space allotted for the content or the overarching categories that specific items fall into, but also the potential rhetorical effect of the description. Again, because no a priori categories were used, citations from the speech text itself function as evidence. After the careful analysis of each speech, the last section of this chapter provides a summary of findings and an answer to RQ 1.

Right Livelihood Awards Acceptance Speech

On December 9, 1984, Wangari Maathai accepted the Right Livelihood Award in Stockholm, Sweden. She shared the award with Imane Khalifeh for her peace efforts in Lebanon, Ela Bhatt and the Self-Employed Women's Association in India, and Winefreda Geonzon of the Free Legal Assistance Association (FREE LAVA) in the Philippines. Though shared, it was the first major international award for Maathai or the Green Belt Movement. Although the GBM had been active since 1977 and had begun to draw the attention of some international donors in recent years, the organization was still operating in relative obscurity (Maathai, 2007). This award and the acceptance speech that followed provided Maathai with the opportunity to introduce a larger group of people to the Green Belt Movement. As such, the analysis of Maathai's acceptance speech gives insight into the workings of the movement as represented by Maathai's rhetoric.

Background. Often referred to as the "Alternative Nobel Prize" (Plon, 2005), the Right Livelihood Awards were founded in 1980 by journalist Jakob von Uexkull to

> recognize the efforts of those who are tackling [the challenges now facing humanity] more directly, coming up with practical answers to challenges like the pollution of our air, soil and water, the danger of nuclear war, the abuse of basic human rights, the destitution and misery of the poor and the over-consumption and spiritual poverty of the wealthy. ("Right Livelihood," para. 2)

Realizing that efforts to meet contemporary challenges often did not fit into preconceived categories, von Uexkull was frustrated with the Nobel Prize Foundation's limitations (Plon,

2005) and wanted to provide a forum that would publicize and laud these efforts. The concept of "right livelihood" is appropriate for this award because it refers to the fifth fold of Buddhism, which teaches that individuals need to find a way to make a living without doing harm to others (Lopez, 2001). It also advances the notion that each individual is responsible for his or her actions and their impact on this planet. In that vein, the Right Livelihood Awards stress practical solutions to global and personal problems.

By 1985, the Award was so renowned that the annual award ceremony began to be hosted in the Swedish Parliament. Although recipients have received a steadily increasing monetary prize (Plon, 2005), the true value of the award is in the publicity that follows. As recipients and their organizations are exposed to foreign donors that might not otherwise have heard of them, there is a potential to increase not only monetary donations but also international awareness of their cause. Thus, the acceptance speech provides the unique opportunity to introduce the audience to the person's cause and organization. Maathai's 1984 speech took advantage of this chance by outlining the GBM's structure and activities.

Content and structure. Because this speech represented one of the first opportunities to expose the international audience to the Green Belt Movement, Maathai focused on explaining the GBM's composition and structure as well as its activities. As mentioned above, the Right Livelihood Awards provide speakers and their organizations with the opportunity to reach out to potential donors. As such, Maathai chose to discuss elements that would introduce the activities of the GBM in a way that would entice the audience to consider supporting the GBM in the future. With that in mind, Maathai addressed 11 different areas that potential donors might find of interest.

Maathai first set up the idea that although *she* had officially been awarded the Right Livelihood Awards, she was not the only one who was involved in the project. In fact, she contended that she was accepting the award "on behalf of..." a wide variety of people (Maathai, 1984, para. 1), and acknowledged the importance of everyone involved in the GBM. This sharing of acclaim showed the audience that while Maathai might be the public face of the GBM, she was only one small part in a larger effort, an important notion for the development of the frame in the speech.

Next, Maathai commented on what the GBM does. This section functioned as a preview to the speech. Not only did Maathai describe the GBM's basic activity of tree planting, but she also provided some basic background information on the philosophical underpinnings of the organization. Once the audience understood that the basic activity of tree planting was designed to reverse the adverse effects on the environment caused by human activity, they could better anticipate the main argument.

In the next part of the speech, Maathai listed the reasons for the GBM's involvement in tree planting. In addition to providing specifics on the GBM, this section contributed significantly to the frame development. Maathai explained the reasons for creating the GBM and narrated the development of the organization, the people involved, and their responsibilities. After addressing the GBM's development, Maathai returned to the original idea of this section and concluded by listing the major reasons for the tree-planting program: "One of the most obvious results of deforestation and bush clearing is soil erosion" (1984, para. 23), which "has precipitated an energy crisis because wood fuel has become scarce" (para. 24), which, in turn, "precipitates another problem: malnutrition" (para. 25). This section constitutes the majority of the speech, having itself 10 numbered subpoints, indicating its importance to the overall frame of the speech. Clearly, telling the story of the GBM introduced the audience both to what the group does and to why it does what it does. At the same time, however, describing the hardship allowed Maathai to build identification of her audience with Kenyan women: although most of them had not had to face living without access to sufficient water, fuel, food, or shelter, they could relate to this existential fear. For many of them, the first instinct was likely to be a desire to help.

Of particular interest in this section of the speech is the last paragraph. While Maathai seemingly was discussing just another area of GBM activity in this section, it did not fit with the straightforward descriptions of the rest of the speech. Here, Maathai suggested that one aspect of the GBM's involvement had been the promotion of a positive self-image for women. Curiously, unlike in the other sections, she did not lay out specific elements of the GBM that had contributed to this area of interest, nor did she describe how or why the GBM had become involved. Instead, she used this moment to describe the public standing of women in Kenyan politics as well as the scorn women encountered if they condemned the limitations on their civic participation. Considering the length of the paragraph, this subject was clearly important to her, although it was not tied directly into the rest of the speech. As the following framing analysis suggests, this paragraph seemed to be the beginning of a frame that was not yet fully thought-out.

Starting with next part of the speech, Maathai addressed several pragmatic areas to prove the GBM's success and impact to the audience, such as the organization's short- and long-term objectives (1984, para. 27), the origin of the GBM's funding (para. 51), the use of the funding (para. 53), and the organization's cooperation with the Kenyan government (para. 60). Interspersed with these more pragmatic areas, however, were answers to more philosophical questions. For example, Maathai addressed why she believed that this approach had worked so far, while answering the question, "[W]hy did it take women to start the green belt movement?"

(para. 57). All of these sections were again designed to provide the international audience with enough information about the GBM to entice potential donors and/or supporters to become invested in the GBM's success long after Maathai had finished this speech.

Finally, Maathai concluded by addressing the inevitable question, "[W]hat of the future?" (1984, para. 63). She was intent on demonstrating to the audience that the financial award would be put to good use. She achieved this objective by explaining the importance of every human being's involvement in changing the current course of environmental destruction and emphasized the GBM's desire to reverse that course as well as its intention to use the financial award to that end. By stressing the GBM's clear plan of action for using the monetary award, Maathai implied to the audience that its potential donations would be put to good use as well. While she never specifically asked the audience for donations or support, the topic selection and organization of the speech suggest that Maathai considered this moment an opportunity to advance the GBM's cause as well as its approach to social and environmental change.

Frame development. In her Right Livelihood Awards acceptance speech, Maathai framed the possibilities for social and environmental change in terms of interconnectedness. While she never used the term itself throughout the speech, the selection of the topics discussed and the salience placed on certain themes suggest that Maathai wanted her audience to understand the importance of interconnectedness to social and environmental change. Maathai realized that simply introducing the audience to the GBM would not result in lasting change. Framing allows the rhetor to provide the audience with a different perspective or interpretation of an issue, subject, or event; thus, Maathai attempted to provide her audience with a viable solution to the environmental and social problems facing the planet. For her, this solution starts by recognizing the importance of interconnectedness.

To advance her argument, Maathai relied on two claims: (1) understanding the impact of cascading effects and (2) the importance of a global/local approach. Throughout the speech, Maathai developed each of these elements by making certain themes more salient. These themes then became the building blocks for the frame.

Cascading effects. As mentioned above, the vast majority of the speech revolved around the third section, where Maathai laid out the GBM's organization, activities, and development. She used this section to document the impact of cascading effects, both negative and positive. To advance her frame of interconnectedness, Maathai first needed to make the audience understand that each action results in something else, leading to a trickling effect. She did this by focusing on two themes: (1) the negative effects of cutting down trees indiscriminately and (2) the positive impact of the GBM's mundane act of planting trees.

Maathai stressed the importance of understanding the cascading effects when she recited the commitment declaration used by the GBM before planting trees:

> Being aware that Kenya is being threatened by the expansion of desert-like conditions, that desertification comes as a result of misuse of the land by indiscriminate cutting-down of trees, bush clearing and consequent soil erosion by the elements; and that these actions result in drought, malnutrition, famine, and death, WE RESOLVE to save our land by averting this same desertification by tree planting wherever possible. (1984, para. 7, emphasis in original)

This pledge explains clearly that a cascading effect is possible from a single negative action: Cutting down trees indiscriminately leads to soil erosion, which leads to drought and in turn to malnutrition, famine, and death. Although the emphasis of the pledge is on the effects of negative actions, it focuses on the possibilities that open up if the GBM continues to fight against it by planting trees. To this day, reciting this declaration is a ritual at each GBM tree-planting ceremony because it reinforces the interconnectedness of all things, positive and negative, to its members (Maathai, 2007). Realizing that one action can have such a detrimental impact, while another has the potential to reverse that effect, is the first step in understanding the interconnectedness frame.

Maathai furthered the theme of the negative effects of cutting down trees indiscriminately when she later explained in more detail how these elements are related. The way she described the effects of deforestation allowed the audience to follow the steps in the chain:

> One of the most obvious results of deforestation and bush clearing is the soil erosion. During the rainy seasons rivers are red with the top soil. Lost top soil leaves behind impoverished sub-soil which cannot support agriculture and as a result food production goes down....
> Deforestation and bush clearing has precipitated an energy crisis because wood fuel has become scarce. Fetching of wood and preparation of food for the family is a responsibility of the women. And so as wood disappears women and children walk further and further from home to look for firewood which may only turn out to be twigs and sticks. Where these do not exist they will turn to agricultural residue and cow dung. These are products which should be returned to the soil in order to make it richer for food production. Burning these breaks the carbon cycle and creates a vicious cycle in agricultural production.
> The crisis of wood fuel precipitates another problem: malnutrition. A woman with little wood fuel opts to give her family food that requires little energy to prepare. If she has money she often turns to refined foods like bread, maize meal, tea and soft drinks. A woman may not appreciate what she must give her family to ensure a balanced diet. That ignorance, coupled with shortage of wood fuel provides an excellent background for undernourishment and diseases associated with poor feeding habits. If too many people are caught up in this situation one can easily have a sick society and a sick society is unproductive. Unproductive people are eventually pushed down into the world of underdevelopment. (1984, para. 23, 24, & 25)

While this quotation sounds like a laundry list of items, Maathai drove home the point that the problem of cutting down original trees went beyond mere environmental destruction; it had hindered the development of the people and had led to nearly insurmountable setbacks. By spending a significant portion of the speech outlining this negative cascading effect, Maathai impressed upon her audience that, while some simple act such as cutting down a tree might seem relatively unimportant in the grand scheme of things, understanding that this one simple act was irrevocably connected to everything else was crucial in achieving social and environmental change.

At the same time, Maathai demonstrated the interconnectedness in the positive actions that the GBM had engaged in, beginning with the mundane act of planting trees. She focused much of the third part of the speech on telling the story of how the GBM had developed and functioned. It had begun by taking trees to the people, who "clamored for the trees we issued at public meetings" (1984, para. 12). Then the group had realized that the people needed to relearn how to grow trees because they had gotten used to cash crops and forgotten more traditional methods of farming. Once this had been achieved, "the demand for tress necessitated the establishment of tree nurseries" (para. 9), which resulted in "the need to train ordinary persons to become seedling producers" (para. 9) and the decision "to make rural women groups our major target groups" (para. 9). It also led to the decision "to purchase seedlings at a minimal price of about US seven cents per seedling. This way not only do the groups gain new and useful knowledge but tree production becomes income generating." (para. 14). Again, she illustrated the point that one action leads to another, even within the approach taken by the GBM. Each part of the GBM's effort was connected to another, both in the negative impact the organization was trying to overcome and in the proactive measures it used to solve its problems.

This notion was well demonstrated when Maathai talked about the people involved in the organization in addition to women's groups. As she stated, "[T]he original major objective of the green belt movement was to help the needy urban poor of a certain area of Nairobi. In mind were the handicapped, school leavers, and the very poor" (1984, para. 17). As Maathai explained, the solution was to hire them as green belt rangers and nursery attendants who after "basic training would [be] able to nurse the trees and assist the school children each of whom attends a few trees" (para. 17). According to Maathai, this not only provided people with income who otherwise would have had difficulties finding employment, but it also expanded the workings of the GBM. The nursery attendants helped schoolchildren tend to trees; the schoolchildren, in turn, learned how to take care of the environment. The process thus generated awareness of the benefits of indigenous trees to those could share that knowledge in the future. Furthermore, Maathai elaborated, "[W]e noted that when the community identifies [a] person who could play this role they would mostly identify a very poor parent whose children may be having problems with school fees" (para. 18). As a result, Maathai explained, this additional income then allowed these parents to keep their children in school, advancing their education and possibly improving their future.

Interestingly, while the pledge recited by the GBM focuses mostly on the negative cascading effects of cutting down trees, in her speech Maathai reversed the order and focused first on the GBM's positive impact. In terms of basic speech organization, this structure makes a lot of sense. Most students of rhetoric learn the principles of primacy and recency in their basic speech class (e.g., Beebe & Beebe, 2009; Griffin, 2009): The principle of primacy suggests that audiences will best remember what they hear first, while the principle of recency supposes that they better remember the last thing they hear. The least effective placement of arguments is in the middle because audiences tend to recall arguments placed in the middle the least. While Maathai established both positive and negative cascading effects as building blocks for the interconnectedness frame, she wanted the audience to focus on the possibilities for social and environmental change. Because she believed acknowledging interconnectedness would provide a potential solution to the problems facing humanity, Maathai wanted the audience to see the effectiveness of the GBM's particular approach so that it would be willing to accept the frame she offered. Maathai achieved that goal by placing the GBM's positive impact first, as well as spending more time on this subject than on anything else in the speech.

Global/local approach. The second claim Maathai advanced in her speech was the importance of a global/local approach. If things were truly as connected as she suggested, then change would depend on the involvement of everyone. By arguing for a global/local approach, Maathai suggested that the impact of the GBM's actions would be limited if the rest of the world did not approach change in a similar fashion. Again, she relied on two themes to stress her argument: (1) the need for community at the local as well as the global level, and (2) the need to accept responsibility at both the global and the local level.

The first theme for this claim can be found in the introduction, where Maathai emphasized the importance of community. As previously mentioned, she took this moment to highlight the idea that even though she was accepting the Right Livelihood Award, she was only accepting it for all of those involved in the GBM's efforts:

> I have not only come on my own behalf, but on behalf of the National Council of Women of Kenya (NCWK), especially the numerous women groups who produce the tree seedlings in the fifty odd tree nurseries, the thousands of school children who plant them and take care of them under the dedicated leadership of their teachers. I have come on behalf of the green belt staff who give a presence of the movement in remote places of our country, the individuals who have planted trees on their plots and on behalf of any person who has sponsored a tree in any of the green belts. I have also come on behalf of the donors who gave us funds to be able to translate our ideas into a programme. (1984, para. 1)

In this introduction, Maathai suggested that this community functions at the local level in Kenya as well as the global level. The GBM's achievements to date had only been possible because of all the people involved, including farmers in Kenya and donors abroad. Without the financial support from outside of Kenya, the GBM never would have made it this far. In terms of framing the approach to social and environmental change as interconnectedness, the Maathai wanted the audience to see that its role in change was as important as that of the women and farmers in Kenya who were actually planting the trees.

Maathai emphasized the importance of community not only by mentioning all of the people who had won this award with her, but also through her language choices. Throughout the speech, Maathai utilized first-person plural pronouns, "we" and "us," to argue that change had to be a combined effort. The only times she referred to herself were to express gratitude in the first two as well as the last two sentences. Apart from that, she used the first-person singular "I" only when she addressed the question, "[W]hy did it take women to start the green belt movement?" (1984, para. 57). Yet, even in this part of the speech, she deferred to others who had helped turn into reality the idea she "was just lucky" (para. 58) to have had. As she detailed, "I think that women in the NCWK were quite good at pursuing that idea which for a long time bore little fruit" (para. 59). Again, the argument focused on community and the willingness to work together rather than individual success. By reiterating this notion of community, Maathai allowed the audience to feel welcome to contribute to the GBM's cause. The repetition of the argument increased the salience of this particular argument and frame for the audience.

The second building block for arguing for a global/local approach was advanced during the introduction, when Maathai mentioned that the GBM had been "telling [people] that with a little bit of help from outside and much will to use their resources on their part they can reverse the trend" of soil erosion, drought, malnutrition, and famine (1984, para. 2). This suggested that a global/local approach would depend on the acceptance of personal responsibility on both levels: local and global.

Support for this idea can be found in Maathai's explanation of why the GBM was actively planting and growing trees:

> Both bodies [the Kenyan Ministry of Environment and Natural Resources, and the Presidential Commission on Soil Conservation and Re-Afforestation] are responsible for re-afforestation efforts in the whole country. But we know that few governments, and less so in the developing world, can afford the financial and man-power resources required to do what needs to be done. It is necessary for private/voluntary, non-governmental organizations (NGOs) and individuals to be mobilized to provide at least the man-power needed in afforestation programs. (1984, para. 11)

While it would probably have been easier to simply blame the government for not doing its job, Maathai argued that because everything is interconnected, we all need to be responsible for our own actions. In terms of the GBM, that meant Kenyans had to realize that they needed to become actively involved in reversing the trends of soil erosion and all its effects by planting trees. For the audience, it meant that they need to become active by support the GBM's efforts and accepting its approach to social change.

Although the need for personal responsibility was interspersed throughout the rest of the speech, it was particularly emphasized in the last section, where Maathai discussed the future use of the monetary award she was receiving. She began by speaking of personal responsibility in more general terms to make the audience understand that it belonged not only to Kenyans, but to all of them as well. As Maathai contended,

> We must continue to care and bother about issues which are not immediately concerned with the gratification of our physical senses. We are a unique heritage to the ecosystem on this planet earth and we have a special responsibility. If to those to whom more has been given more will be expected then we must embrace our special responsibility which is more than is expected of the elephants and the butterflies. In making sure that they and their future generations survive we shall be ensuring the survival of our own species. (1984, para. 64)

Thus, according to this argument, for change to occur, all of humanity needed to realize that everything is interconnected and can only be changed if we all feel responsible for taking action, even if it does not seem to impact us immediately. This in turn necessitated an approach to change that did not focus only on one area or geographic location, but combined the local with the global.

To emphasize this idea, Maathai concluded her speech by talking about the how the GBM would contribute to change beyond Kenya. As she explained, "Kenya is not among the worst [in terms of environmental destruction] in Africa. And so we must go beyond Kenya and help raise awareness in other parts of the Continent" (1984, para. 65). Maathai explained that the GBM planned to do just that by using the money of the Right Livelihood Award to "establish a trust which could be used to provide seed money for the establishment of programs similar to the green belt movement elsewhere in Africa" (para. 66). Indeed, in 1986, the Pan-African Green Belt Network was established, and Maathai and the GBM supported workshops teaching leaders from other nations the basics of the GBM's approach (Maathai, 2007). To facilitate these teachings, Maathai published a manual that explained the philosophy, development, and structure of the GBM (Maathai, 1985 & 2004a). By advancing the mission of the GBM beyond Kenya into other African nations, the GBM actively advances a local approach, demonstrating the importance of feeling responsible for social change beyond one's personal geographic location.

Gender inequality argument. As mentioned above, the last paragraph of the third section seems slightly out of context with the overall framing efforts of the speech. In this section Maathai focused on the standing of women in Kenyan politics as well as the negative reactions to those who attempted to improve the political participation of women. The length of the paragraph suggests that this subject was clearly important to Maathai, but she did not spend enough time on it over the course of the speech to warrant an independent frame. Instead, it seems to be something that she herself was still articulating.

Although Maathai did not explicitly state what the GBM had done to diminish the unequal treatment of women or to promote their positive self-image, the fact of the matter is that the GBM is predominantly a women's organization, founded by a women's group, and employs mostly women. As such, improving the life of women has always been a central concern for Maathai and the GBM. It stands to reason that this section of the speech should hint at a future frame of inequality; however, Maathai seemed reluctant to develop it further in this speech. One possible explanation for this is that at the time, she considered it more important to focus on the interconnectedness frame to advance the GBM's mission of social and environmental change without losing too many supporters. As an outspoken woman, Maathai had experienced firsthand how drawing attention to gender inequality can adversely affect one's professional and personal life (Maathai, 2007). It is therefore possible that, although she considered this matter important, she was not certain that a deeper discussion would not alienate her audience and, ultimately, detract from the main goal of introducing the GBM.

By concluding the speech talking about the GBM's commitment to bringing its approach to other nations, Maathai tied both her central claims together: Because she and the GBM recognized the cascading impacts of positive and negative actions, the organization was willing and dedicated to a global/local approach. As such, framing social and environmental change in terms of interconnectedness had the desired effect: Once people understood that each action has consequences, Maathai implied, it would become self-evident that they must also be proactive and think about these consequences beyond their own horizons. Seeing this interconnectedness would open up possibilities otherwise not acknowledged: Instead of being perceived as an exercise in futility, tree-planting would comes to be perceived as a real possibility for enacting lasting change.

"The Bottom is Heavy Too: Even with the Green Belt Movement" – The Fifth Edinburgh Medal Address

In April 1993, Maathai accepted the Edinburgh Medal, an award given to an individual who has not only proved to be an exceptional scientist or technologist, but one who has also contributed to the social well-being of the community in which he or she works (P. Hymers, personal communication, December 13, 2010). Maathai accepted the award during a time of great personal and political turmoil. In the early to mid-1990s, Maathai was heavily involved in democratization efforts in Kenya, and both she and the Green Belt Movement found themselves attacked repeatedly (Worthington, 2003). These attacks ranged from slanderous media campaigns, to physical expulsion from the government-owned offices in Nairobi, to near-fatal personal attacks against Maathai (Maathai, 2007). In term of the GBM's activities, this time represented a turning point as the organization moved from merely planting trees to actively engaging in political campaigning. As such, Maathai's speech can shed light on the framing of the GBM at this crucial time.

Background. The Edinburgh Medal was first awarded in 1989 in conjunction with the first Edinburgh International Science Festival (P. Hymers, personal communication, December 13, 2010). This festival was the first of its kind in the world, celebrating science and technology by encouraging people of all ages to discover the wonders of the world around them ("About Us," n.d). The medal is one of the highlights of the festival, and is given to a person who not only has distinguished him- or herself in the sciences, but whose contributions have made a lasting impact on the community. It is this added emphasis on contributions to the social well-being of the scientist's community that distinguishes the Edinburgh Medal from other science awards. As Ramphal (1994) explained, "[T]his is not a mini Nobel Prize for Science; it is more like a mini Nobel Prize for Peace limited to scientists" (para, 3).

Maathai received that year's medal because of her application of scientific knowledge in a community that needed practical, employable solutions. In his introduction to the address, Lord Provost Norman Irons (1994) explained that Maathai's "scientific, her environmental and her social activities have combined together in a manner that has brought benefit to hundreds of thousands and in which she has persevered under both administrative and physical threat" (para. 1).

Content and structure. Unlike Maathai's Right Livelihood Award speech, this acceptance speech did not function exclusively to introduce the Green Belt Movement. In fact, even the award itself seemed rather unimportant. While Maathai referenced her involvement in science as well as its importance, it was not the central topic of the speech. Instead, Maathai focused on describing the challenges facing those at "the bottom of the pyramid," as she called them. Ultimately, this speech was a scathing critique of the circumstances that result in poverty from environmental degradation and highlighted the GBM as a possible solution.

The other major difference from the Right Livelihood Award acceptance speech was the structure. While the previous speech had had clearly labeled sections, this one did not. Instead, Maathai relied on a simple problem/cause/solution pattern. According to Griffin (2009), rhetors choose this organizational pattern for two primary reasons: first, if the speaker believes he or she will be more persuasive by explaining how the problem came about; and second, if the rhetor believes that describing the causes can help the audience see the merits of the proposed solution. That Maathai spent the majority of the speech discussing the causes suggests that she believed her speech would have a greater impact on the audience if they were exposed to the causes of poverty resulting from environmental degradation and, consequently, might be more willing to accept the GBM as a possible solution.

Maathai began the speech by explaining what she considered to be the problem. She argued that the purpose of people's lives is to strive for happiness and fulfillment, regardless of their living situations. This is an idea the audience could easily relate to. But, as she pointed out, most people at the bottom face serious hurdles in attaining this goal and, more importantly, often they cannot even meet their basic needs. As such, Maathai attempted to draw parallels and differences between her audience and the bottom of the pyramid so that the audience could better identify with frustrations resulting from the challenges the bottom faces.

Maathai identified these challenges as the cause of the problem. She divided them into three major areas: unequal knowledge of science, environmental degradation, and the difference between childhood dreams and reality. The first hurdle Maathai mentioned was unequal knowledge of science. She touched on this subject when she discussed the purpose of life as a journey toward happiness and fulfillment to suggest that science had contributed to significantly improving the global quality of life. Considering that Maathai was accepting a science award, it stands to reason that her audience shared this belief. Yet, Maathai asserted there were several problems with this advancement brought on by science that her audience might not fully comprehend. First, although humans better understood the world, they still did not take good care of it. More importantly, however, the benefits of science had been distributed unevenly. According to Maathai, while those at the top could take advantage of scientific advancement, those at the bottom still viewed it as magical and out of reach.

The uneven distribution of scientific knowledge also plays a role in the second hurdle Maathai mentioned: environmental abuse. She contended that the people both at the bottom and at the top negatively influence the environment: the former from lack of understanding, the latter from greed. Again, Maathai used this section to draw a clear distinction between the bottom of the pyramid and the top. While Maathai never said that the audience was part of the top, it was implied.

The third hurdle Maathai described was the difference "between reality and childhood dreams" (1994, para. 12). This was the longest and most involved topic discussed in the speech. To illuminate the importance of this aspect, Maathai reached out to her audience by explaining that many people are brought up to believe that a good education allows the fulfillment of their dreams. This again was an idea that her audience could identify with because they themselves had probably been brought up in that belief or impressed the importance of education onto their own children. But Maathai pointed out that, unlike her audience, those at the bottom then encounter man-made obstacles that "prevent them from utilising much of the knowledge, expertise and the experience they have acquired in their studies and in the course of their lives" (para. 12). To illustrate her point, Maathai described own experience: She had graduated from college and received an appointment at the University of Nairobi, where she then faced discrimination because of her gender. Maathai continued with this example of her experience, recounting the challenges she had encountered in her marriage because she was educated and wanted to continue her work. By telling her own story, Maathai moved the discussion from the abstract to the specific. Being able to associate a face with the story made it easier for the audience to identify with her and, by extension, with the subject of discussion.

Focusing on her own story also provided Maathai with a perfect segue into the last section of her speech, where she presented the GBM as a solution. She stated that her research had led her "into environmental activism" (1994, para. 20) and with that the creation of the

GBM. In terms of audience effect, it was important for Maathai to provide a solution at this point because her description of the challenges facing the bottom had likely caused significant discomfort for her audience. As mentioned, Maathai had engaged in identification-building strategies, and as such, the audience would want to know what could be done to alleviate the problems of those at the bottom. By describing the approach of the GBM, she provided the audience with a possible out. They could reduce their discomfort by engaging in some of the principles she suggested and by supporting the GBM and similar organizations.

Maathai concluded the speech by impressing upon her audience the importance of understanding the disastrous impact the sheer number of the bottom has on the planet. She also returned briefly to the idea that science might provide a solution to the problems outlined above. But ultimately, she suggested to her audience that science would be ineffective as long humans did not change their ways.

Frame development. Overall, the content and structure of the speech already suggest that Maathai's primary concern at this point did not lie with the interconnectedness frame previously advanced in the first speech. Instead, she argued that systemic inequality perpetrated by the top on the bottom was the root cause for the problems facing the Third World. Framing systemic inequality as the cause, she framed interconnectedness as a possible solution. As Entman (1993) argued, one of the features of framing is to define problems, causes, and solutions. While the previous speech had focused mostly on framing the solution, this speech concentrated predominantly on the causes of poverty as a consequence of environmental degradation.

Inequality. Maathai relied on two basic claims to develop the inequality frame: (1) that the bottom contributes to its own challenges, and (2) that the top prolongs existing inequality.

Focusing mainly on the contributions of each group allowed Maathai to argue for a systemic frame: If inequality is deeply rooted in both groups, then it permeates every aspect of life. Discussing the bottom's part in addition to the top's role might also have made her audience feel less attacked and, by extension, might have resulted in their acceptance of her message. Again, each of the claims was supported by a number of salient themes that were incorporated throughout the speech.

As detailed above, Maathai used the majority of her speech to outline inequality as the root cause for poverty and environmental destruction by describing in detail the hurdles people at the bottom face in their struggle for happiness and contentment. In explaining these hurdles, Maathai also argued that certain realities of the bottom contribute to the continued inequality they experience. It is important to note, though, that while Maathai employed this strategy to balance the attacks against the top, she did not necessarily blame the bottom. Three themes are of particular importance for this element of the frame: (a) a lack of understanding, (b) the perpetuation of the myth of easy advancement, and (c) the continuation of certain traditional norms and values.

While perhaps not the most significant theme discussed in the speech, lack of understanding was nonetheless important to Maathai. She focused this element primarily on environmental degradation and the bottom's role in it. Maathai argued that

> [t]he resources on the planet earth are not only limited, they are also being degraded. The people at the bottom of the pyramid do not understand limits to growth and they do not appreciate that as they seek their own happiness and fulfilment they could adversely affect the same resources and jeopardise the capacity of future generations to meet their own needs. (1994, para. 8)

For those at the bottom, the lack of understanding meant that they continued to use resources unwisely, without thinking about the consequences of living in a closed system. Not only had the lack of understanding resulted in unwise use of resources, it had also led those at the bottom to be misguided in their approach to enacting change. Because the "majority of the people at the bottom of the pyramid are both the causes and the symptoms of environmental degradation" (para. 10), they often do not see that their long-term behavior change needs to be part of the solution. Instead, Maathai argued, "[t]he majority of the people at the bottom of the pyramid would rather deal with the symptoms because their objectives are short-term and are directed towards immediate survival" (para. 20). Repeatedly emphasizing this lack of understanding throughout the speech, Maathai suggested that it significantly inhibits people's ability to move beyond inequality. By not seeing their own behavior as problematic, they are more likely to dismiss ideas that would require serious change on their part. As such, lack of understanding is a building block for the inequality frame.

The second theme Maathai made salient with her claim that the bottom's realities contribute to the inequality frame is the perpetuation of the myth of easy advancement. As mentioned earlier, the idea of education as a means to advance one's status was an idea that the audience could easily appreciate. Many of them had been told this as children themselves and would tell their children the same story. For Maathai, the idea that education is a savior was a myth, a fact that became apparent whenever someone attempted to make it a reality. Maathai contended,

> [m]any of us at the bottom make our children believe that education is the key to a good job, a good salary and a good quality of life. They believe that education will get them out of the bottom of the pyramid and provide comfort without effort. It seems easy enough because passing examinations and moving to the next grade may come easily. As they struggle through school they console themselves with the promised success which will ensure them a place at the top of the pyramid. If that depended on good grades and certificates many of us would have little to worry about. We would be on the top! (1994, para 11.)

While a good education was certainly not the cure-all at the top either, it was much less so at the bottom. Maathai argued that in reality the myth of easy advancement was just another element put in place by the system to reinforce the inequalities faced by the bottom. Once school was over and grades were earned, people found that the reality was vastly different. As she explained,

> [b]etween reality and childhood dreams are many man-made hurdles which the people at the bottom fight against all their lives as they try to overcome them and to achieve meaningful development, improve their quality of life and realise full potential. These obstacles prevent them from utilising much of the knowledge, expertise and the experience they have acquired in their studies and in the course of their lives. This knowledge and experience is supposed to make the journey surer and easier. But there is a big difference between childhood dreams and the reality of the pyramid. At the bottom of the pyramid, sooner or later we all learn that. (para. 12)

Maathai implied that by continuing this myth, the people at the bottom set up their children for failure. Instead of focusing on being able to put their expertise to good use, the youth still believed in the easy fix of education and were disappointed and ill-prepared when easy advancement did not happen. To emphasize her point, Maathai used her own life story. This allowed the audience to put a face with the idea and, as such, made it easier for them to empathize with the problems faced by the bottom.

Maathai's story, however, was mostly designed to delineate the last theme for this claim: that the continuation of some traditions and norms contributes to persistent inequality because it creates varying degrees of inequality even among those at the bottom. For instance, traditions and norms establish codes of conduct for those at the bottom that are designed to keep everyone in their assigned places. These traditional expectations create almost insurmountable obstacles for women. This section of the speech was given the most space and elaboration, indicating the importance of this element of inequality to Maathai. As her life story shows, women face inequality from all directions. Even if they succeed in obtaining a good education and a well-paying job, they have little hope of advancing beyond their initial position. As Maathai detailed,

[m]obility upwards was too slow. It was as if I did not matter as much as the others. There was something I did not have and I could not have. The hurdle had nothing to do with passing examinations, having certificates or being a good teacher. It had everything to do with my gender! (1994, para. 16)

Gender was a factor that for Maathai far surpassed other elements contributing to the inequality frame. In her experiences, it was a compounding element that was at work even amongst those at the bottom, designed to create a hierarchy of the bottom. Maathai remembered,

Several years later I was in the village of my birth and childhood and I was at home with people who were black like me. I was still not o.k. This time tough, it was my gender that was the problem. I have since learned that at the bottom of the pyramid there are very strict cultural and religious norms which govern the birth, life and death of women in society. These age-old traditions make the bottom quite heavy. (para. 17)

That Maathai chose to point out how traditions and norms had worked to put her back into her traditional and stereotypical gender role was particularly surprising considering that she knew the audience saw before them a successful scientist and social activist who had just received an important science award. However, it would also have reminded them of the impact of continued traditions and norms on people's ability to advance their lives. Chances are that members of the audience had experienced similar situations, such as women who had been told they could not be scientists, or men with a working-class background who had been told they had no business attending college, or any of a number of other possibilities. While Maathai tied this theme to the contributions of the bottom's realities to the inequality frame, it would have been easy for the audience to make the leap to investigate how their own circumstances might have influenced them, leading in turn to their increased appreciation for the fight of the bottom against inequality.

Although Maathai spent much of her time talking about the contributions of the bottom to inequality, understanding the impact of the top was more crucial for the speech. Because this speech took place in Edinburgh, Scotland, it is important to remember that the audience was most likely made up of dignitaries, scientists, Scots, and others who would have been part of what Maathai called the top. With regard to the second frame advanced in the speech, it was necessary for that audience to realize their impact on inequality: only through this recognition would they accept interconnectedness as a possible solution to social and environmental problems. While Maathai needed to be forceful enough for her audience to get the message, she also needed to keep her argument abstract enough to avoid alienating them.

Maathai achieved the latter portion of this requirement by talking about the top in very abstract terms. She never referred to her actual audience as members of the top, nor did she ever personally address the audience using the second-personal singular pronoun "you." Instead, she kept the "top" vague and faceless. At the same time, she used two strongly worded themes to support her claim that the top was at least equally to blame as the bottom, if not more so. The two themes she relied on were (a) the greed of the top and (b) their willful ignorance.

Maathai suggested that the top contributes to the plight of the bottom by being greedy. Although Maathai used the word "greed" in her speech to address how humanity's greed impacts the environment, she did not use the word itself to explain how the top's greed leads to inequality. As such, she never told the audience per se that their greed was a significant part of the problem. Instead, she implied it by incorporating the idea into her discussion about science and technology. She mentioned that "commercialised science has greatly enriched societies which have made scientific discoveries and have been able to apply them and create new and efficient tools" (1994, para. 7) and wondered,

> [W]ill those who have this know-how be willing to share it when it gives them the advantage over the bottom? How can they when with that advantage they (the top) can exploit not only their own resources but also the untapped resources belonging to those at the bottom of the pyramid? (para. 7)

Thus, science in the hands of the top contributes to inequality: It allows those who understand the world better to take advantage of this knowledge. For example, she explained,

> the top is blinded by insatiable appetites backed by scientific knowledge, industrial advancement, the need to acquire, accumulate and overconsume.... It promotes the lifestyle of those at the top of the pyramid and sells it as the ultimate in fulfilment and happiness" (para. 8)

This then creates an ever-increasing schism between the haves and have-nots. While Maathai did not go into detail about the lifestyle people were striving for, the audience could likely have identified exactly what it was they could do or have because of science and technology. And although she did not ask them specifically whether they would be willing to give up their own lifestyle to help balance the world, the question was implied.

At the same time, Maathai argued that those at the top are unable or unwilling to see that their ignorance can have serious consequences for the entire world. "For as long as we sustain a pyramid the bottom will continue to gather momentum and may take all of us with it where it is always going... the abyss of the bottom" (1994, para. 25). Although she did not provide as much detail for this theme, it runs as an undercurrent throughout the speech.

Looking at the topics discussed in the speech, it becomes apparent that Maathai felt the need to make explicit certain notions the audience might not have comprehended otherwise. For example, she began the speech by explaining how all people strive for happiness and fulfilment. This, again, was an idea that the audience could easily identify with. As she suggested, "[W]e wake up every morning to toil on the resources available to us so that we can realize the goal of happiness and fulfilment" (1994, para. 2). She continued by pointing out explicitly that this goal is unattainable for the bottom. And not only are happiness and fulfilment impossible to attain, but "there are not enough resources to meet even our basic needs" (para. 2). Her decision to be this explicit suggests that Maathai believed the top to be ignorant of the realities the bottom was facing.

For Maathai, the challenge with the top's ignorance becomes apparent when the top is trying to help. Maathai maintained that

> [m]any governments, aid agencies and charitable organisations invest heavily in the symptoms of environmental degradation as they mop up the world. Less effort and enthusiasm is demonstrated in dealing with the causes of the garbage they are so willing to mop up. (1994, para. 10)

Because the top focuses on the symptoms rather than the causes of inequality, they are likely to build schools, suggest education as a solution, and bring in food aid and cash crops. While Maathai did not mention any of the above specifically, her topic choices alluded to this. Although she did not fault the top explicitly for the myth of easy advancement through education, she went to great length to explain to them why education does not allow advancement for the bottom. The simple fact that she related her own struggles in epic detail suggests that she feared the top still could not understand. It was this ignorance, she believed, that prevented the top from accepting measures that might tip the balance and eliminate the pyramid.

Interconnectedness as a solution. In addition to establishing inequality as the root cause for poverty and environmental destruction, Maathai framed appreciation for interconnectedness as a first step toward a solution. While less dominant in this speech, the notion of interconnectedness Maathai had already established in her Right Livelihood Awards acceptance speech was reflected here as well. She reiterated the idea that "environmental degradation is brought about by soil erosion, deforestation, pollution and loss of biological diversity in our earth systems" (1994, para. 9). But she did not go to great length to establish what interconnectedness is. Instead, Maathai set up the frame of interconnectedness as part of a solution by (1) explicitly calling for a shift in thinking and (2) using the GBM as a practical example.

That a shift in thinking needed to occur was a notion Maathai advanced early in the speech. As soon as she had demonstrated that the life goal of happiness and fulfillment was impossible to achieve for the vast majority of people on the planet, she contemplated that

> more and more people now strongly suggest that we are one species which needs to be less arrogant and exploitative against what St. Francis called our brothers and sisters in the wide spectrum of creation. Every other species has a right to exist and to pursue its own happiness and fulfilment and has no obligation to homo sapiens. The species should assist each other and help each other to achieve the goal of happiness and fulfilment. Homo sapiens, by virtue of its higher intelligence and a capacity for love and compassion should be more a custodian and less the exterminator. (1994, para. 6)

While Maathai here was referring more to our interaction with nature than with one another, the notions behind this part of the speech suggest that her solution required a major shift in thinking, not only with regard to how we treat one another but how we treat our planet. Recognizing how all living things on the planet are connected with one another, she implied, would have to be a major factor in achieving any kind of meaningful change, because it was greed that was depleting the resources available.

More importantly, Maathai aligned understanding the importance of interconnectedness with her primary frame of systemic inequality. According to Maathai,

> [soil erosion, deforestation, pollution and loss of biological diversity] in turn are brought about by political and economic policies and activities which are dictated by greed, corruption, incompetence and an insatiable desire to satisfy the inflated egos and ambitions of those who wield political and economic power. They are exacerbated by population pressure, international debts and interest rates, low prices for export goods, commodity protectionism and inevitable poverty. (1994, para. 9)

Because the interconnectedness of all things exacerbates the problems caused by the systemic inequality, recognizing that interconnectedness is the first step toward a solution. Maathai wanted the audience to understand that simply implementing temporary fixes for random problems would not get to heart of the issue. Instead, Maathai needed them to realize that at the center was recognition of interconnectedness. She suggested that

> [p]erhaps part of the answer lies with man itself. Humans have to reassess their roles on this planet, reassess their values, reassess their understanding of the universe and perception of what constitutes their happiness and fulfilment. We may have to reassess our system of governance and seek security and peace not in a pyramid but in a balanced and harmonious whole. (para. 25)

Clearly, she was urging her audience to reconsider their attitude and asking them to engage in a shift in thinking that would place the utmost importance on interconnectedness.

To help her audience understand what it would take, Maathai described the GBM as a practical example. Just as telling her personal story helped create comprehension about the myth of education, the GBM's story made the solution more tangible. When talking about the work of the GBM, Maathai explained that one of the fundamental differences between the GBM and other organizations was "[t]he overall objective of the Movement... to raise awareness of symptoms of environmental degradation and raise the consciousness of people to a level that would move them to participate in the restoration and the healing of the environment" (1994, para. 20). Focusing on the foundational principles of interconnectedness, Maathai argued that the GBM began its work by advocating for a shift in thinking. Appreciating the interconnectedness of the environment with people's living situation was crucial to create effective change.

It is important to note, though, that for Maathai, recognition of interconnectedness was just a first step in overcoming the problems of poverty and environmental degradation. Once the recognition was there, action needed to be taken that would support this primary argument and provide practical solutions. Describing the importance of the GBM's tree-planting program for the women, she explained, "Without education, capital, political and economic policies to support them they find themselves engulfed in vicious cycles of debilitating poverty, lost self confidence and never-ending struggle to meet most basic needs" (1994, para. 22). While it was important to understand systemic man-made inequality as the root cause and interconnectedness as a first step toward a solution, Maathai also reminded her audience that much more needed to be done: Rethinking how they did things might just allow them to effect actual change.

Nobel Peace Prize Lecture

On October 8, 2004, the Nobel Peace Prize committee announced that Wangari Maathai of the Green Belt Movement in Kenya was that year's Peace Prize laureate "for her contribution to sustainable development, democracy and peace" ("Press Release," 2004, October 8). She accepted the award on December 10, 2004, in Oslo, Norway. As was customary, she gave the Nobel Lecture in conjunction with the receipt of the Nobel diploma and medal. Because this represents the highest honor Maathai and the GBM have received to date, the speech was chosen as the last to be used for the framing analysis in this chapter. It provides insight into Maathai's framing of social change and the Green Belt Movement during this important phase of the organization.

Background. Based on the last will of scientist Alfred Nobel written in 1895, five Nobel Prizes are awarded annually ("Nobel Prizes," 2011). Nobel left much of his wealth for the establishment of a foundation designed to award a diploma, a medal, and a cash award for significant achievements in each of the following five categories: medicine, physics, chemistry, literature, and peace. The Nobel Peace Prize is awarded "to the person who shall have done the most or the best work for fraternity between nations, the abolition or reduction of standing armies and for the holding and promotion of peace congresses" (Nobel, as cited in "Nobel Peace Prize," 2011). Each year a separate Nobel Peace Prize committee requests nominations from qualified nominators and, after a thorough review process, announces the year's laureate at the beginning of October. Although the other four prizes are administered by the Nobel Foundation in Nobel's native country of Sweden, he directed that the Peace Prize be given out by a

Norwegian committee. As such, the Nobel Peace Prize is the last prize to be awarded during Nobel week each year and, unlike the other awards, the ceremony takes place at City Hall in Oslo, Norway. Since its inception in 1901, the Nobel Peace Prize has become one of the most important and prestigious peace prizes in the world (Lundestad, 2001) and has been given for work in a wide range of arenas, including humanitarian efforts and peace movements, advocacy of human rights, mediation of international conflicts, and arms control ("Nobel Peace Prize," 2011).

Like the laureates before and after her, Maathai was chosen from a competitive field of nominees because she was considered to best embody the founding principles outlined by Nobel. As the committee explained in its press release,

> Maathai combines science, social commitment and active politics. More than simply protecting the existing environment, her strategy is to secure and strengthen the very basis for ecologically sustainable development. She founded the Green Belt Movement where, for nearly thirty years, she has mobilized poor women to plant 30 million trees... Through education, family planning, nutrition and the fight against corruption, the Green Belt Movement has paved the way for development at grass-root level. We believe that Maathai is a strong voice speaking for the best forces in Africa to promote peace and good living conditions on that continent. ("Press release," 2004, October 8)

In terms of prestige, the Nobel Peace Prize is certainly one of the highest honors anyone could be awarded. Additionally, Maathai was the first and to date the only African woman to have been awarded this prize. As such, it stands to reason that Maathai's lecture was highly anticipated. As with the Right Livelihood Awards, Maathai's immediate audience was made up of potential donors and supporters. In addition, because the Nobel Peace Prize is one of the most prestigious awards on the planet, and the speech took place during the new media age in 2004, it was safe to assume that this speech would be broadcast and reported across the globe. It provided Maathai with a unique opportunity to share the GBM's success and approach with a larger-than-usual audience.

Content and structure. Much like the first speech covered in this chapter, Maathai's Nobel Peace lecture followed a clear structure. Yet, in contrast to her Right Livelihood Awards speech, Maathai did not number her various points in this lecture. This time, she created divisions between the different portions of the speech by re-addressing her audience. The phrase "Excellencies, ladies and gentlemen" (2004b, para. 1, 10, 26, 41) became a signifier to indicate that she was about to change gears. Each of the four sections covered a particular content area: introduction/thanks, the Green Belt Movement and its development, call-to-action, and conclusion.

The introduction of the speech served two purposes. First, Maathai fulfilled the requirements of an acceptance speech by expressing her gratitude for the award and explaining the meaning of this award to her and the GBM. Second, this introduction set the tone for the rest of the speech by providing a preview of the other two major sections. The second section explained how the work of the GBM was tied particularly to women's rights, education, and democracy efforts, while the third section was a call-to-action for the international community as well as African leadership.

Maathai's second section addressed the GBM's work and its development from mere tree planting to democracy efforts. This part was important because it described the GBM to the audience, providing them with a clear picture of Maathai's vision for social and environmental change. As previously mentioned, it is safe to assume that Maathai's audience was much more significant than usual and, as such, provided her with the rare opportunity to persuade people from across the globe. Consequently, this section needed to draw specific connections between the various elements of the GBM's work. Thus, Maathai not only touched on the cornerstone of the GBM – tree planting – but connected it to the importance of women to the organization. She focused on their role in African societies: "[W]omen are the primary caretakers, holding significant responsibility for tilling the land and feeding their families. As a result, they are often the first to become aware of environmental damage as resources become scarce and incapable of sustaining their families" (2004b, para. 12). This painted the picture of women as first defenders of the environment. Continuing her story about the development of the GBM, Maathai turned to the idea of civic education as a way to help farmers make the connections between the environment, their current living situations, and their political standing. From education, Maathai then moved to democratic engagement. According to Maathai, the GBM had expanded its activities to include democratic efforts when it became clear that effective stewardship of the environment would not be possible without democratic space.

Interspersed with the idea of democratic engagement, Maathai also addressed the destruction of cultural heritage and its impact. This was the only one of the three speeches under investigation that considered cultural heritage as a separate issue. So far, she had referred to traditional norms and values as contributing factors in inequality. In this speech, however, she focused on the positive impact of culture on the environment.

She signaled the beginning of the third section of the speech by once again addressing the audience with the phrase "Excellencies, ladies and gentlemen" (2004b, para 26). This time, however, Maathai also used the device to subdivide the section itself to refocus the audience's attention. Overall, this section was a call-to-action: Maathai asked the audience to become involved in environmental efforts and peace-building. But she attended to different sections of her audience and emphasized different types of expectations for involvement. Maathai first focused on the more general need to re-engage with the environment and pointed to the importance of the involvement of humanity itself in this endeavor. Then, she turned to African leaders and specifically called on them to act:

> I call on leaders, especially from Africa, to expand democratic space and build fair and just societies that allow the creativity and energy of their citizens to flourish. Those of us who have been privileged to receive education, skills, and experiences and even power must be role models for the next generation of leadership. (2004b, para. 31, 32)

As mentioned above, the fact that Maathai was the only African woman to have won this award was of particular importance, not only to her but also to the African continent. It meant that the world was taking note of the work happening in Africa. Additionally, part of Maathai's notoriety was due to her continued engagement in Kenya's democracy efforts (Maathai, 2007). By singling out this section of her audience, she reminded African leaders that, although great strides had been made, their work toward democracy and peace was not done. The time and space spent on this group suggests that getting her point across was of particular importance to Maathai.

She also separated this portion of the third section from the remainder by using the address "ladies and gentlemen" (2004b, para. 35). Although Maathai used this signifier at this point in the speech, it was not designed to set up a new topic. Instead, it was meant to draw the audience's attention away from Maathai's appeal to African leaders and back to the global audience. Maathai had started the third section by appealing to the more general involvement of everyone in environmental and peace efforts; now she returned to this segment of the audience but provided various groups with more specific ideas for their involvement.

Maathai concluded the speech with a story about a stream that had run by her house when she was a little girl. She reminisced,

> I would drink water straight from the stream. Playing among the arrowroot leaves I tried in vain to pick up the strands of frogs' eggs, believing they were beads. But every time I put my little fingers under them they would break. Later, I saw thousands of tadpoles:

black, energetic and wriggling through the clear water against the background of the brown earth. (2004b, para. 42)

Although separate from the previous section, this conclusion continued the call-for-action as Maathai told her audience that "the stream has dried up" (para. 43) and that "the challenge is to restore the home of the tadpoles and give back to our children a world of beauty and wonder" (para. 43). Ending with such a heartwarming, naturalistic story drove home the urgency of the GBM's work as well as the need for global involvement.

Frame development. Exposed to her largest and most diverse audience yet, Maathai used this opportunity to propose the GBM's "holistic approach to development" (2004b, para. 40) as a solution to environmental and social problems facing the planet. As such, she set up the frame that environmental and social change requires a holistic solution. While the term *holistic* most commonly refers to the medical field and its attempts to deal with the whole person instead of just treating the physical condition, the term originates from *holism*, the notion that nature produces wholes and not parts ("Holistic," 1989). Keeping both definitions in mind, a holistic approach to change requires addressing the problems and causes simultaneously. Furthermore, it entails viewing potential solutions as needing to fit with the wholes provided by nature.

Maathai suggested that the GBM's approach did so through two key approaches: (1) the interconnectedness of all actions and (2) the need for a global/local approach. As in the previous speeches, Maathai developed these claims by making certain themes more salient than others. In this speech, however, she wove all of the elements together to establish the frame, using the metaphor of the tree, as will be demonstrated toward the end of this section.

Appreciation of interconnectedness. Although Maathai returned to the idea of interconnectedness more strongly in this speech than in the previous one, appreciating interconnectedness served only as a supporting claim in this speech, not as a frame itself. The two supporting themes she made salient here were (a) the raising of awareness and (b) the impact of culture.

Much of the speech focused on describing the development of the GBM in terms of how its activities had aided in raising its members' awareness about interconnectedness. Maathai illuminated for her audience how various elements of human action and interaction are connected not only to the environment, but to other aspects of social change, such as women's rights, sustainable development, good governance, and democracy. Raising people's awareness of these connections is a crucial element in the GBM's approach. It allows members to see the whole picture and take individual actions accordingly. At the same time, Maathai introduced the audience to the GBM's various activities and its history so they could also develop a new appreciation for interconnectedness. By showing how this cornerstone of the GBM's approach had worked successfully for the organization and its members, Maathai demonstrated to her audience that a holistic approach was the right course of action beyond Kenya.

The importance of raising awareness became apparent when Maathai traced how the GBM had moved from tree planting to democracy efforts. For Maathai, women's rights are the natural extension of tree-planting, civic education the natural extension of women's rights, and democracy the natural extension of civic education. Starting with addressing the "lack of firewood, clean drinking water, balanced diets, shelter and income" (2004b, para. 11), the GBM was automatically pointed to the need of including all of these other social concerns as well. For Maathai, then, social change can happen only when the notion that "sustainable development, democracy and peace are indivisible" (para. 8) is recognized. This recognition is possible only when people are aware of the interconnectedness of the various elements.

Maathai followed this line of thinking when she talked about the GBM's move from mere tree planting to its involvement in civic education programs. The challenge the GBM faced was not only that the women did not know the importance of tree planting or even how to plant trees, but that they also did not feel responsible enough to take action. As she explained,

> [i]nitially, the work was difficult because historically our people have been persuaded to believe that because they are poor, they lack not only capital, but also knowledge and skills to address their challenges. Instead they are conditioned to believe that solutions to their problems must come from 'outside'. Further, women did not realize that meeting their needs depended on their environment being healthy and well managed. They were also unaware that a degraded environment leads to a scramble for scarce resources and may culminate in poverty and even conflict. (2004b, para. 16)

Thus, before the GBM could facilitate effective change, the women needed to become aware of the interconnectedness of deforestation, soil erosion, drought, inequitable development, and their part in this development. At the same time, Maathai suggested that the women had needed to become conscious of social and political issues beyond the environment. Civic education classes had remedied the situation. As Maathai argued,

> [W]e developed a citizen education program, during which people identify their problems, the causes and possible solutions. They then make connections between their own personal actions and the problems they witness in the environment and in society. They learn that our world is confronted with a litany of woes: corruption, violence against women and children, disruption and breakdown of families, and disintegration of cultures and communities. (para. 17)

Over the course of the classes, the women were able to air their problems and figure out how to tackle them. More importantly, Maathai suggested that the community had made the necessary connections to become more politically engaged as well. She clarified that

> [i]n the process, the participants discover that they must be part of the solutions. They realize their hidden potential and are empowered to overcome inertia and take action. They come to recognize that they are the primary custodians and beneficiaries of the environment that sustains them. Entire communities also come to understand that while it is necessary to hold their governments accountable, it is equally important that in their own relationships with each other, they exemplify the leadership values they wish to see in their own leaders, namely justice, integrity and trust. (para. 19, 20)

By raising people's awareness of the interconnectedness between actions and impacts as well as actions and possibilities, Maathai argued, the GBM was improving people's willingness to become more engaged with the environment as well as the political and social processes around them. Thus, raising awareness was the first building block to establish interconnectedness as an element of a holistic approach to social change.

Another crucially important element for Maathai was culture and the restoration of cultural heritage. While not new itself, this idea had not been articulated to this degree in the previous speeches used for this analysis. Though Maathai had argued for traditional farming methods and the benefits of indigenous trees, she had not yet made cultural heritage a separate theme to advance interconnectedness. Furthermore, during her Edinburgh Medal address, Maathai had raged against traditional norms and values that heightened various levels of inequality amongst those at the bottom, impacting women in particular. In this speech, however, she elevated restoration of cultural heritage as "the missing link in the development of Africa" (2004b, para. 33). She argued that by rediscovering and accepting the positive elements of their culture, people "would give themselves a sense of belonging, identity, and self-confidence" (para. 34). If one remembers the definitions of *holistic*, this idea is reminiscent of the medical field's desire to treat the whole person and not merely the physical condition. In terms of framing the GBM's holistic approach as the solution to environmental and social problems facing the world, raising awareness addressed the physical condition, while cultural restoration added to treating the whole.

To provide the audience with a specific example, Maathai described the African tradition of using thigi trees to encourage reconciliation. In particular, she contended,

> [u]sing trees as a symbol of peace is in keeping with a widespread African tradition. For example, the elders of the Kikuyu carried a staff from the *thigi* tree that, when placed between two disputing sides, caused them to stop fighting and seek reconciliation. Many communities in Africa have these traditions. (2004b, para. 22)

It was these traditions, Maathai suggested, that needed to be remembered and restored, because "[s]uch practices are part of an extensive cultural heritage, which contributes both to the conservation of habitats and to cultures of peace" (para. 23). This explicitly reminded the audience that cultural heritage was an important element in appreciating interconnectedness: It is cultural heritage that teaches people about the interconnectedness between themselves and nature, while also reminding them of practices that encourage peace and reconciliation. Broaching this topic with the mostly non-African audience was crucial in advancing her vision of a holistic approach: While she did not explicitly charge the so-called developed world with undermining and ridiculing African traditions, she did suggest that the spread of Western, or new, values was having a detrimental effect on other cultures. More specifically, she said that "[w]ith the destruction of these cultures and the introduction of new values, local biodiversity is no longer valued or protected and as a result, it is quickly degraded and disappears" (para. 23). While Maathai explicated the environmental repercussions of losing biodiversity, she implied that the loss of cultural heritage would also lead to dissolution of important societal values. Appreciating how culture was connected to the environment and the social fabric of societies was vital in achieving any meaningful change. Restoring this cultural diversity was part of the GBM's holistic approach. Specifically, she contended, "[f]or this reason, the Green Belt Movement explores the concept of cultural biodiversity, especially with respect to indigenous seeds and medicinal plants" (para. 23).

It is important to note, though, that although Maathai delineated the restoration of cultural heritage as an important building block for the appreciation of interconnectedness, she did not suggest that all cultural elements needed to be preserved. Unlike her previous speech, Maathai's Nobel lecture did not outline specific aspects of culture that discriminate against women or other groups. But she mentioned that "[c]ulture is dynamic and evolves over time, consciously discarding retrogressive traditions, like female genital mutilation, and embracing aspects that are good and useful" (2004b, para. 33). By drawing this distinction, she indicated to her audience that she was aware of the obstacles some traditions bring with them. At the same time, she encouraged her audience to look beyond these more challenging traditions and appreciate the positive impact they could have on development.

Global/local approach. The second claim Maathai used to advance the frame that environmental and social change require a holistic solution focused on implementing this agenda across the globe. While Maathai outlined the work necessary on the local level when she described the GBM's development, she spent much of the speech describing in detail what was expected of (1) Africans and (2) the international community. Maathai's concentration on these two groups did not indicate that the local level was of less importance, but it did imply that she believed more work had yet to be done in other arenas. People at the local level usually are much more accepting of the GBM's program than political leadership or the far-removed international community, simply because they can see the benefits.

During her introduction, Maathai first suggested that she would highlight the need for Africans to take an important role in this holistic approach. While Maathai outwardly addressed her "fellow Africans" (2004b, para. 7) as a whole, her language implied that she was calling on called on African leaders specifically to do their part in advancing democracy. She called upon them "to intensify our commitment to *our* people, to reduce conflicts and poverty and thereby improve *their* quality of life... [to] embrace democratic governance, protect human rights and protect our environment" (para. 7, emphasis added). This notion was confirmed later in the speech when Maathai outlined the involvement needed to effect change. She stated, "I call on leaders, especially from Africa, to expand democratic space and build fair and just societies that allow creativity and energy of their citizens to flourish" (para. 31).

Although she did not spend significant time on developing this theme, it nonetheless became clear that outlining the responsibility of Africans and African leadership was of particular importance to her in this speech: They were the only individual group she mentioned in addition to several elements of the international community. As suggested earlier, one of her reasons for highlighting the importance of Africans might have been her distinction as the first African woman to win the Nobel Peace Prize. This award drew the world's attention not only to Kenya in particular, but to the African continent in general. As such, she wanted the world to see that Africans could and would play an integral part in their own development. The fact that she seemed to call on African leaders specifically could have been the result of her own political fights over the years. While there is no specific evidence for this in the text, Maathai's efforts for democracy in Kenya had been long-standing (Maathai, 2007). Because she had had first-hand experience fighting against dictatorship, she may have wanted to use this stage to call on other African leaders to continue their commitment to democratic political systems.

Yet, it would not have been enough if Africans were the only ones to implement a holistic approach to change. If the so-called developed world did not change its approach to development, peace, and the environment, no amount of change in Africa was going to make much of a difference. For that reason, Maathai appealed most strongly to the international community and the rest of the world to accept a holistic approach. A first indication of this required change came during the introduction, when Maathai mentioned,

> In this year's prize, the Norwegian Nobel Committee has placed the critical issue of environment and its linkage to democracy and peace before the world. For their visionary action, I am profoundly grateful. Recognizing that sustainable development, democracy and peace are indivisible is an idea whose time has come. (2004b, para. 8)

Not only did she reiterate the importance of interconnectedness in this instance, but more importantly, she pointed out that most people had not yet recognized this connection. By thanking the committee for "their visionary action" (para. 8), Maathai suggested that viewing environmentalism, peace, and democracy as inextricably connected was still a revolutionary idea. This notion was supported by the reactions that followed the announcement of Maathai's nomination. Many observers were perplexed by the committee's choice that year, calling the Peace Prize a farce because they did not understand how tree planting could facilitate peace (Bethell, 2004/2005). In mentioning the committee's extraordinary decision, Maathai acknowledged that lack of understanding and used the moment to call on her audience to participate in the acceptance and practice of the GBM's approach.

Maathai also suggested that the international community needed to accept its responsibilities. She first hinted at this element when she recounted the challenges the GBM faced in the beginning, in stating, "[O]ur people have been persuaded to believe that because they are poor, they lack not only capital, but also knowledge and skills…. Instead they are conditioned to believe that solutions to their problems must come from 'outside'" (2004b, para. 16). While Maathai did not explicitly fault the international community for this conditioning, she did imply it. As such, she spent much of the third section of the speech outlining what the international community must do to advance a holistic approach to environmental and social change.

Maathai left no doubt that the international community had to become engaged for the GBM's approach to be successful outside of Kenya. Neither did she leave any room for hesitation:

> In the course of history, there comes a time when humanity is called to shift to a new level of consciousness, to reach a higher moral ground. A time when we have to shed our fear and give hope to each other.
> That time is now.
> The Norwegian Nobel Committee has challenged the world to broaden the understanding of peace: there can be no peace without equitable development; and there can be no development without sustainable management of the environment in a democratic and peaceful space. This shift is an idea whose time has come. (2004b, para. 28, 29, 30)

Not only did Maathai reiterate the importance of a holistic approach that appreciates interconnectedness, but she also added a sense of urgency. After calling on African leaders, Maathai listed what various elements of society needed to do in order to achieve peace. It was this part of the speech that provided that audience with a possible response to Maathai's request for change. Not only did she ask "governments to recognize the role of social movements" (para. 36), "civil society [to] embrace not only their rights but also their responsibilities" (para. 36), "industry and global institutions [to] appreciate that ensuring economic justice, equity and ecological integrity are of greater value than profits at any cost" (para. 37), and "young people to commit themselves to activities that contribute toward achieving their long-term dreams" (para. 39); Maathai also reminded her immediate audience that "[t]he extreme global inequities and prevailing consumption patterns continue at the expense of the environment and peaceful co-existence. The choice is ours" (para. 38). To get to the higher moral ground, Maathai contended, everyone must change their thinking about the environment, governance, development, consumption, rights, and much more. As mentioned before, if peace, equability, and social change were to be a reality, her audience needed to accept that "sustainable development, democracy and peace are indivisible" (para. 8) and take a "holistic approach to development" (para. 40). This holistic approach could only be achieved if they accepted interconnectedness across the globe and acted accordingly.

The tree. Connecting all of these elements was the symbol of the tree. Even during the introduction, Maathai had already used the metaphor of "seeds of peace" (2004b, para. 4) when she explained the meaning of having won the Nobel Peace Prize:

> Although this prize comes to me, it acknowledges the work of countless individuals and groups across the globe. They work quietly and often without recognition to protect the environment, promote democracy, defend human rights and ensure equality between women and men. By so doing, they plant seeds of peace. (para. 4)

Throughout the remainder of the speech, the tree became the metaphor that wove together interconnectedness and the need for global change to establish the frame that a holistic approach was the road to environmental and social change.

Maathai picked up the metaphor of the tree as a signifier for interconnectedness in the second section of the speech by talking about the GBM's tree-planting activity. At first, "[t]ree planting became a natural choice to address some of the initial basic needs identified by women" (2004b, para. 14). But the tree planting also expanded women's rights because the "women gain some degree of power over their lives, especially their social and economic position and relevance in the family" (para. 15). It was also the tree that led the GBM to offer civic education courses because the women did not understand how the trees would help their situation. By understanding the interconnectedness of things, the women of the GBM began to rethink themselves, and the process of shifting began. Lastly, the tree connected the GBM's democracy efforts and culture to women's rights, education, and environmentalism because

> [t]he tree became a symbol for the democratic struggle in Kenya. Citizens were mobilised to challenge widespread abuses of power, corruption and environmental mismanagement. In Nairobi 's Uhuru Park, at Freedom Corner, and in many parts of the country, trees of peace were planted to demand the release of prisoners of conscience and a peaceful transition to democracy (para. 21)

During the struggle for democracy, the GBM not only planted peace trees but used these trees to settle conflicts, remembering the aforementioned tradition of thigi trees.

Although Maathai did not explicitly link this metaphor to the second claim used to advance a holistic frame – global involvement – she subtly suggested the literal importance of trees to the environment in her conclusion about the stream of the tadpoles: the stream had dried up because of deforestation. Maathai implied that if the international community began to work holistically against environmental destruction and social inequality, then there was the chance that this home of the tadpoles could be restored and that we could "give back to our children a world of beauty and wonder" (2004b, para. 43). It is clear from this analysis that for Maathai, social change can come only when we begin to shift our thinking and understand the importance of a holistic approach to social and environmental change. For Maathai, all of this can happen through something as simple and innocuous as planting a tree.

Summary of Findings

Presented with the opportunity of addressing international audiences filled with policy makers, potential donors, and supporters, Maathai used her speeches to effectively frame both the need for environmental and social change, and the GBM's approach to the topic. As the above analysis has shown, each speech offers significant insight into Maathai's framing strategies. Thus, following the chronological timeline of the rhetorical acts furthers the understanding of the frame development over the course of the movement's existence.

Because Maathai's Right Livelihood Awards acceptance speech was the first major international award for Maathai or the movement, Maathai had to use the speech not only to introduce the workings of the GBM, but also to bring forth a first attempt at framing its approach. The analysis suggests that Maathai concentrated on developing an interconnectedness frame: that is, she argued that understanding the interconnectedness of actions and effects is a crucial step in solving environmental and social problems facing the planet.

To advance this frame, Maathai relied on two claims, each supported by several themes represented in the speech. The first claim Maathai made was that positive and negative actions both have cascading effects. By describing the GBM's activities and the people involved in the movement, as well as the trickle effect of environmental destruction, Maathai chose to discuss topics that would illuminate clearly how intricately connected each action is to the next. The second claim used to support the interconnectedness frame tied cascading effects to a global/local approach. The audience needed to understand that each of its actions has consequences, not only for them but across the globe: that everyone needs to become engaged in environmental and social change, and that we are all interconnected with one another.

During the Edinburgh Medal address, Maathai took a different approach. While she still framed interconnectedness as part of the solution, the majority of the speech focused on framing inequality as the root cause for environmental and social problems. This meant that, in contrast to the previous speech, the second speech under investigation relied on two frames, each of them supported by a number of claims that can be derived from the themes advanced throughout the speech.

The more significant frame of this speech was concerned with establishing inequality as the root cause for the problems of environmental destruction and poverty. Maathai delineated this frame by making both the bottom's and the top's contributions to inequality salient. While she posited that the top's contributions were steeped in greed and willful ignorance, she suggested that the bottom's contributions were more attributable to their realities than to their willing participation. Acknowledging the roles of both groups in inequality allowed Maathai to argue for the importance of accepting interconnectedness as a crucial element in finding a lasting solution. It should be noted that Maathai outlined the bottom's impact of creating several layers of inequality and, with them, a hierarchy of the bottom. The most important of these layers was discrimination against women, which prevented any meaningful advances of women in much of the Third World.

The second frame Maathai established in this speech was the understanding of interconnectedness as the potential solution to inequality. As compared with her Right Livelihood Awards acceptance speech, Maathai did not spend the majority of the argument explaining interconnectedness. Instead, she focused her supporting claims on explicitly calling for a shift in thinking at both the global and local levels. To demonstrate how this shift in thinking could facilitate interconnectedness as a solution, she used the GBM as a practical example.

Lastly, in the Nobel Peace Prize lecture, Maathai's framing of the GBM shifted gears again. While inequality as a frame had been particularly significant in the second speech under investigation, Maathai focused this speech on establishing the frame that environmental and social change requires a holistic approach like the one used by the GBM. In terms of the supporting claims, this speech is quite interesting: Maathai relegated the frame of interconnectedness established in the previous two speeches to a supporting argument, while she also returned more strongly to the idea of a global/local plan of attack.

In arguing for taking a holistic approach to change, Maathai used interconnectedness as a cornerstone argument to cement her frame. If people understood how intricately connected every aspect of the world is, then, Maathai suggested, our solutions to any problem would be far more effective. To support this claim, Maathai relied on two major themes throughout the speech.

First, she acknowledged that the biggest issue with her vision was that people had not yet fully appreciated the importance of interconnectedness and, with that, the need for a holistic approach. As such, her supporting theme focused on the need to raise awareness, describing the GBM's activities as proof of the effectiveness of raising awareness. Second, for the first time Maathai explicitly mentioned the importance of restoring cultural heritage as an element of interconnectedness. This theme became important when people started appreciating the new way of thinking as proposed by Maathai. While she asked her audience to increase their awareness of interconnectedness, she also wanted to ensure that important traditional wisdom did not get lost and societies did not lose their heritage and identity.

In addition to interconnectedness, Maathai explained that a holistic approach requires the involvement of everyone at the global and local level. While she used the GBM's work as an example of how this approach looks at the local level, she supported her argument by focusing thematically on the international community and African leaders. Both of these groups, she suggested, still needed convincing. For this approach to be successful, they would need to recognize their role in it.

This analysis suggests that the frame(s) employed by Maathai changed over the course of the 20 years under investigation. Yet, although there was a shift, much of the content and themes remained the same. What changed, primarily, was the emphasis Maathai placed on various elements. Looking at each of the speeches individually has provided insight into Maathai's framing efforts.

CHAPTER 4 – PEDAGOGICAL ANALYSIS

Pedagogical Analysis of the Green Belt Movement

The analysis in this chapter answers RQ2 and investigates whether the GBM's practices follow critical pedagogical principles. For that purpose, two primary texts were used: (1) *The Green Belt Movement*, a booklet describing the GBM's work and organization, written by Maathai in 1985; and (2) *The Green Belt Movement: Sharing the approach and the experience*, an expanded version of the earlier manual written by Maathai and published in 2004. Each text provides valuable insight into the GBM's practices through descriptions of the organization, activities, and procedures. Using both manuals provides insight into the development of the GBM's educational practices over time.

Despite the fact that both manuals were authored by Maathai rather than a variety of GBM members, this circumstance actually provides consistency not always available for the rhetoric of social movement organizations. As already explained, Maathai has been the public face of the GBM since its inception and, because of her deep immersion in the organization, has shaped public perception of the GBM (Ndegwa, 1992; Worthington, 2003). More specifically, these manuals were designed to share the GBM's approach with a non-Kenyan audience. Over the past 30 years, that audience's perception of the GBM has been shaped predominantly by Maathai. Thus, instead of being problematic, Maathai's authorship of these manuals proves beneficial.

This chapter is divided into three sections. In the first section, the practices found in the 1985 manual are discussed; the second section focuses on the differences between the 1985 edition and the 2004 expanded edition; and the third section provides a summary of findings about the GBM's pedagogical approach. The first two sections each begin with a brief description of the books and their place in the GBM's development, followed by an analysis of the practices described in each text. For this purpose, this chapter relies on the tenets outlined by Paulo Freire (1970), who argued that conscientization, praxis, and dialogue are all necessary elements for a critical pedagogical approach. As previously mentioned, critical pedagogy is a teaching philosophy intended to empower the marginalized and disenfranchised by focusing on the student's lived experience. According to Freire (1970), critical pedagogy involves three primary elements: conscientization, praxis, and dialogue. Conscientization is the process of raising students' critical awareness about their placement in the world. Praxis is the tool that helps in the conscientization process by making knowledge meaningful and applicable to the student's life. Lastly, dialogue needs to occur between all parties involved to create knowledge. Exploring the manuals for conscientization, praxis, and dialogue provides an answer to RQ2. While each of these three elements is investigated separately, they are intricately connected with one another, and for critical pedagogy to occur, all three must be present.

The Green Belt Movement (1985)

Background and content. Published in 1985, this booklet of 77 pages came in the wake of Maathai's and the GBM's first international award, the Right Livelihood Award. As mentioned in the previous chapter, the award acceptance speech provided Maathai with the opportunity to introduce the GBM to a largely international audience full of potential donors. One of the points she particularly emphasized in the speech was the GBM's desire to use the monetary award in an effort to expand its approach into other African nations. This booklet seems to be a consequence of that desire. In the preface, Maathai mentioned that GBM members were often approached to discuss the organization and provide field demonstrations and descriptions of the GBM's work. As she contended, "it has now become necessary to document some facts about the movement so that the experience may be more broadly shared with those interested. Hence this booklet" (Maathai, 1985, preface).

Not only was the GBM receiving more international acclaim during the early to mid-1980s, but it also became independent of its founding organization, the NCWK (National Council of Women of Kenya). As Maathai (1985) explained, the GBM's rapid growth led to the decision of registering the movement as a separate society in 1984, with the result that "its aim and objectives can henceforth be pursued independently for the benefit of all" (p. 76). As an independent organization, the GBM needed to ensure its continued growth. One of the ways to spur such growth was to share the GBM's approach with other social movements and social movement organizations.

The development of this manual embodied one of strategies used by the GBM to achieve broader acceptance of its approach. Over the course of eight chapters, Maathai described the GBM's history, objectives, procedures, achievements, finances, constraints, and future. Additionally, she included the forms used by the GBM to establish green belts and nurseries. Much of the same information appeared in several chapters with slightly different emphasis. It appears that the manual served two purposes: (1) It allowed those interested in the GBM's approach to potentially replicate it, and (2) Maathai included just enough information to address questions for those interested in donating to or supporting the movement. This analysis focuses on descriptions of practices and procedures to delineate whether the GBM follows critical pedagogical principles.

Analysis

Intended to explain the GBM's organization and approach to non-members, the booklet gives insight into educational practices used by the GBM to advance their agenda. As previously mentioned, Freire delineates thee elements of critical pedagogy: Conscientization, praxes, and dialogue. Each of these elements is explored in turn.

Conscientization. One of the first steps towards empowerment of the oppressed is the need to increase their consciousness of their own standing. As Freire (1970) argued, "[the oppressed] have no consciousness of themselves as persons or as members of an oppressed class" (p. 46). Thus, to be effective, educational strategies must be developed that awaken people to their current situation. During the time this book was written, the GBM's primary activity was still centered on the tree-planting program. Consequently, much of the conscientization efforts of the GBM focused on increasing people's awareness of their connection to the environment. Additionally, however, an analysis of the booklet suggests that the GBM also worked to develop a positive self-image of its primary constituents and members – women.

One of the GBM's first and foremost goals is to make people aware of their relationship with the environment and the importance of the trees to their lives. While raising environmental consciousness does not seem to be conscientization in the Freirean sense, for Maathai environmental destruction and political oppression are inextricably interconnected. As she outlined during the Right Livelihood Awards acceptance speech, one of the first issues people need to become aware of is the cascading effects of environmental destruction: Deforestation leads to desertification, which leads to a lack of food and water, which in turn results in malnutrition and a lack in productivity (Maathai, 1984). Ultimately then, environmental destruction perpetuates oppression. While Maathai did not explicitly explain the connection between deforestation and oppression in the manual, she implied it by emphasizing the importance of trees in people's lives. In that regard, raising people's environmental consciousness functions as a first step in the conscientization process.

Maathai (1985) proposed community tree planting as a project "to improve settlements and avert desertification" (p. 6). The challenge was to convince people that tree planting could provide the solution to their oppression, because "some members of the committee were not impressed by the idea and some even opposed it. They felt it that it couldn't be done" (p. 6). This required increasing the women's awareness of their own living situation and their ability to improve upon it. In the booklet, Maathai (1985) repeatedly referenced the need to "educate populations on the inter-relationship of environment and other issues" (p. 21), "to promote environmental education" (p. 21-23) and "to promote soil conservation" (p. 23-24). Tree planting had to be explained on a level that was meaningful to the women, which meant that it had to relate to their most pressing needs of food, fuel, shelter, and water:

> It is known that a shortage of fuelwood indirectly promotes malnutrition as women are forced to opt for foods which require little or no energy to prepare, such a rice, maize meal, chapattis, bread and tea.
> When the only available energy sources are maize cobs, maize stocks, sisal stocks, weeds or twigs, women will cook refined rice and maize flour or make tea for children to eat bread with. Where even these sources are not available women have to talk long hours to look for wood or use cowdung, thereby depriving agricultural land of badly-needed manure. (Maathai, 1985, p. 21)

While Maathai did not explicitly state in this section that tree planting was presented as a possible solution to these very real problems, it stands to reason that the GBM would do so. Communicated at that level, the message was simple and useful: by planting trees, the women were told, they would be able to provide their families with income, food, firewood, and clean water. Using this rhetorical strategy, the GBM communicated the message on a meaningful level for the women, and the conscientization process started.

Once the idea took root, the NCWK organized its first community tree-planting ceremony. The following pledge has been recited at each GBM tree-planting ceremony since:

> Being aware that Kenya is being threatened by the expansion of desert-like conditions, that desertification comes as a result of misuse of the land by indiscriminate cutting-down of trees, bush clearing and consequent soil erosion by the elements; and that these actions result in drought, malnutrition, famine, and death; we resolve to save our land by averting this same desertification by tree planting wherever possible. In pronouncing these words, we each make a personal commitment to our country to save it from actions and elements which would deprive present and future generations from reaping the bounty which is the birthright and property of all. (Maathai, 1985, p. 7)

Reciting this pledge at tree-planting ceremonies ensures that the importance of their connection to the environment as well as their ability to change their lives is always at the forefront of members' minds. Additionally, because the ceremonies are open to the public and take place on public land, others from the community are also present (Maathai, 1985). The recitation of the pledge provides the rest of the community with the opportunity to increase their awareness about the environment and their relationship with it. As Freire (1970) posited, "liberating education consist in acts of cognition, not transferrals of information" (p. 79). Attending and actively participating in tree-planting ceremonies provides the opportunity for such acts of cognition to occur.

The GBM's desire to increase people's awareness of their long-term relationship with the environment can also be seen in its consistent involvement in schools. Maathai (1985) recalled that the GBM's second tree-planting ceremony took place at the request of a headmistress at Mary Leakey High School. Since then, much of the GBM's tree plantings have taken place on school grounds. Maathai explained that the GBM hires Green Belt Rangers "who assist the children take care of trees and thereby effectively participate" (p. 63). The students are active participants in the tree plantings as well as in their long-term care. This ensures that they develop an understanding of what it takes to grow trees and the importance of trees to their lives. This in turn means that the children realize at an early age that they themselves have the possibility to change the course of their future by being active participants in it rather than standing by passively.

The second issue Maathai raised in the booklet was the NCWK's and the GBM's desire to improve the image of women. Though it seemed less important during Maathai's Right Livelihood Awards speech, the "development of a positive image of women" (1985, p. 17) was the first short-term goal Maathai mentioned in the book. This prominent placement suggests that increasing women's awareness of their placement in Kenyan society was of the utmost importance. This notion is supported by several other short-term goals, such as "the training of women as cultivators of seedlings" and "to generate income for women" (p. 20).

Achieving an improved image for women might seem more challenging, but because the GBM was founded as a project of Kenya's foremost women's society – the NCWK – the GBM had easy access to its target group. In Kenyan society, women's organizations have always been the driving force behind the social aspects of life (Nzomo, 1989; Udvardy, 1998) and as such could work as a positive force. As Maathai argued,

> the Green Belt Movement, and many other rural projects initiated by women, are exemplary projects not dominated, as men so often claim, by the concerns of the kitchen, babies, nappies or sex. They are good examples of female achievement which should serve at least to encourage women to form a more positive image of themselves. (1985, p. 19)

Tree planting was created as an activity to alleviate the concerns of firewood, food, shelter, and water, all of which are issues women have traditionally been concerned with. But emphasizing the role women played as part of the solution demonstrated that their participation was instrumental in improving their own lives and encouraged the women to rethink themselves. This provided the women of Kenya with an excellent opportunity to engage in conscientization. Additionally, providing women with income through seedling production furthers their positive self-image because it allows them to contribute to their livelihood through less traditional but still acceptable means.

Praxis. As mentioned above, to be considered critical pedagogy, all three elements must be present. In addition to helping with the conscientization process, the GBM had to provide praxis. One of the crucial ideas behind critical pedagogy is making knowledge meaningful and applicable to the student's lived experience. Freire (1970) achieved praxis in his literacy program by first focusing on vocabulary necessary for the peasants' work. The GBM's praxis centers on tree planting. Much of this manual described in detail how the GBM functions, but of particular interest for this analysis are the tree-planting activities themselves and the establishment of the nurseries. Each of these demonstrates how the GBM incorporates praxis.

Clearly, planting trees in and of itself is a practical activity, but that does not yet constitute praxis in a Freirean sense. It becomes praxis when students are actively involved in the creation of knowledge and when this knowledge relates to their lived experience. Freire (1970) argued that if these conditions are met, then the students can engage in reflection of this experience. It is this reflection that turns the object into a subject. Tree planting became meaningful for the GBM members once they could see how the trees positively impacted their lives. The digging of the hole and setting of the tree were all necessary practical activities, but by themselves they were quite meaningless. The women had to make the connection between planting trees and the direct impact the trees would have on their lives in terms of food, firewood, shelter, and water. Once these connections were made, tree planting provided the opportunity for praxis. Additionally, Maathai recounted that at some point the GBM realized the need to

> establish community woodlands… to enhance the beauty of the compounds and create
> windbreaks because we had read of many schools whose roofs had been blown off by the

wind. On several occasions fatal accidents had indeed been reported on such school compounds. By planting trees along the boundaries of compounds we were encircling them with narrow strips or "green belts" of trees" (1985, p. 12)

By increasing the safety of school compounds, tree planting in this instance related immediately to the people's lived experience and moved from insignificance to providing the opportunity for praxis. It allowed the community members to take charge of their lives, moving them from objects to subjects.

Another element of the GBM's practice that facilitated praxis was the establishment of tree nurseries. While these nurseries provide an additional income source for employees, the *establishment itself* constitutes praxis. When the GBM first started, seedlings were provided free of charge by the Department of Forestry (Maathai, 1985). Though the chief conservator of forests "laughed when we told him that we intended to plant fifteen million trees," (p. 11) he still promised the seedlings. Problems arose when he revoked his decision less than a year later. Now the GBM had to pay for the seedlings, using its meager funds for something unplanned. As a result, the GBM founded its own nurseries. It was another opportunity to engage the women in an activity that would diminish their oppression: To gain independence from the governmental nurseries, the women had to rely on their own experiences and knowledge. One of the practices that facilitated the meaningful construction of knowledge was the GBM's decision not to provide nurseries with seeds (Maathai, 1985). Instead, it was suggested that the "farmers collected seeds from their neighborhoods. This method encourages the propagation of trees which are best suited to their neighborhood. Experience has already shown that farmers are very good at deciding which trees will best meet their needs" (p. 68). Although farmers sometimes tend to favor non-indigenous trees that are not as good for the environment, at that stage of the GBM the successful establishment of nurseries was an important step. By relying on trees that would fulfill their immediate needs, the women again took charge of their own lives, moving from object to subject.

Dialogue. The last element of significance for critical pedagogy is dialogue. Dialogue suggests that the oppressed are actively involved in reflective participation; without this reflection and action, Freire contended, "the word is changed into idle chatter, into verbalism, into an alienated and alienating "blah"" (1970, p. 87). Dialogue is a continuously ongoing process, one through which both teachers and students engage in the co-creation of knowledge. For the GBM, dialogue was achieved primarily through the following measures: (1) organizational procedures, (2) training seminars, and (3) community involvement.

The GBM's organizational procedures facilitated the potential for dialogue. As Maathai (1985) explained, although the GBM "has no rigid rules or operation procedures [,] it has developed a set of broad and flexible guidelines… that evolve around the communities involved" (p. 25). While the GBM employs promoters who visit different communities to inform them about the GBM, the initiative for involvement usually has to come from the community itself. Once the community has decided to be engaged with tree planting, it submits an application form to the GBM headquarters. Allowing the communities to take the initiative is an important element for critical pedagogy. It suggests that the GBM does not force anyone to participate, and that it does not simply want to deposit knowledge. Instead, the process requires a desire for critical awareness and engagement. One of the reasons for these guidelines was the pragmatic notion that people would only tend to the trees if they had a vested interest in them. It would not help the environment or the people if the GBM simply planted trees and allowed them to die. If people took the initiative themselves, however, the conscientization process had likely already begun, and by extension the chances for tree survival would be much higher. It also suggests an increased interest in continued conversation and engagement. Once a community or person took the initiative, they would be in regular contact with the GBM staff, which provided the opportunity for engaged learning. It is important to note that conversation itself does not equal dialogue. In fact, Freire (1970) explicitly rejects the notion that dialogue is as simple as conversation. In that regard, placing the responsibility for participation with the communities is of particular importance as the communities have to demonstrate their desire for the co-creation of knowledge.

Throughout the process of establishing a green belt, the GBM provides communities with various checking procedures. Maathai (1985) explained that after applying to the GBM, applicants receive forms with detailed instructions pertaining to the size and number of holes needed for the trees. These must be verified by a staff member before the applicant can retrieve the trees from the nursery. Once the trees have been planted, GBM staff checks on them twice over the course of the following year to ensure survival. While this process seems rather detailed, Maathai contended "the purpose and relevance becomes evident…. because the Green Belt Movement is more than just tree planting. It is an educational experience aimed at changing mental attitudes towards the environment through greater awareness, understanding and appreciation" (p. 26). For conscientization to occur, continuous dialogue is crucial. While the procedures themselves may not constitute true dialogue in the Freirean sense, they begin a conversation that has the potential to develop into dialogue. They provide an opening that allows the women and the GBM staff to engage in shared learning that otherwise would like not occur.

Since the inception of the GBM, one of its cornerstones has been education and training (Maathai, 1985 & 2004a). These training seminars are meant not only to increase people's awareness of their relationship with the environment or the importance of trees to their lives, but also to fulfill the very practical goal of teaching proper tree-planting techniques. As Maathai explained, "most people seem to have little awareness of the care needed between the collecting of the trees from the nursery site and the time when the tree is firmly established in the ground and has become self-reliant" (1985, p. 62). By conducting hands-on training sessions, the GBM not only ensures the survival of the trees, but engages with members in a dialogue that will allow them to create mutual understanding: On the one hand, people learn how to properly care for the trees; on the other, the GBM becomes more aware of the challenges people are facing. As such, the training seminars cover topics such as how to keep livestock away or how to play soccer without destroying the trees. The creation of knowledge that occurs during these training seminars facilitates conscientization by continuously engaging the women in a conversation that allows them to rethink themselves.

Another crucially important way for the GBM to engage in dialogue is community involvement. Tree-planting ceremonies often occur on public land and as such are community acts. Much like the training seminars, tree plantings engage the GBM in dialogue with its members, but the added bonus here is the interaction with the whole community. As Maathai explained, "launching a Green Belt is a big affair…. [It] is usually an opportunity for dancing, friendship-building and general community happiness" (1985, p. 28). One of strategies employed by the GBM to elevate the importance of the event is to invite a guest of honor, "distinguished citizens, senior government officers and residents" (p. 28). This additional layer reminds communities of the importance of reforestation to all levels of society. At the same time, though, it gives the community something to be proud of, increasing the significance of the event for the whole village. As a consequence, Maathai argued, the GBM is usually "requested to return for more tree planting or other development activities" (p. 28), allowing them to continue the dialogue with this particular community.

The Green Belt Movement: Sharing the Approach and the Experience (2004)

Background and content. In 2002, Kenya held its first democratic elections since independence. Although multiparty elections had officially been held in 1992 and 1997, during both of these the Moi regime had tampered with election results, leading to a continuation of the status quo (Nasong'o, 2007). During that time, Maathai and the GBM had become heavily involved in democratization efforts and environmental advocacy (Maathai, 2004a). This led to serious clashes with the Moi regime, resulting in the GBM's expulsion from its original government offices as well as endangering the lives of Maathai and other activists (Maathai, 2007). Yet, change was achieved when the opposition finally united under the banner of the National Alliance Rainbow Coalition (NARC) that was voted into office in 2002 (Chege, 2008). At the same time, Maathai ran successfully for parliament for the Tetu constituency in Nyeri district and was named Assistant Minister for Environment, Natural Resources and Wildlife. Additionally, Maathai was nominated for and awarded the Nobel Peace Prize in 2004 for her work with the GBM.

In recent years, the GBM had undergone significant reorganization through the implementation of a new strategic plan, making it necessary for Maathai to introduce the terms "Phase I" and "Phase II" for the GBM. As Maathai contended, much of "this book is a record of the Phase I experience. It mainly focuses on the activities of GBM from its beginning up to 1999 – the year the strategic planning commenced" (2004a, p. 1). Although the book predominantly discussed the GBM's approach, organization, procedures, and values during Phase I, Maathai outlined anticipated changes.

While Maathai referred to the vastly different political circumstances under which the GBM was now operating, the emphasis of this book was still on sharing the GBM's experiences with others. In the preface she specifically asked her readers to "enjoy this book, learn from it, take the lessons with you and share them." In terms of content, this book covered similar subjects and was divided into similar chapters as the previous edition. One significant difference was the absence of the forms and application letters Maathai had included in her earlier version. Instead, she spent much more time explaining the GBM's "overall goals, values and projects" (2004a, pp. 33-55). A significant addition was a chapter on the "replication of the Green Belt Movement" (pp. 102-110). As an expanded version in both content and time span, this manual provides the ideal tool to continue the preceding analysis of the GBM's educational practices.

Analysis. This section investigates the GBM's practices for conscientization, praxis, and dialogue in the 2004 manual. Exploring this expanded edition provides insight into the development of the GBM's practices over time. Of particular interest for this analysis is an understanding of how the approach has changed and how it has stayed the same since the 1985 manual was released.

Conscientization. In the 1985 manual, Maathai had particularly emphasized the GBM's efforts to raise people's awareness about the importance of appreciating their environment as well as boosting women's positive self-image. While the 2004 book still reflected the same notions, Maathai spent much more time on elaborating consciousness building with regard to people's relationship with the environment. In addition, she described the GBM's efforts in advancing conscientization beyond tree planting.

Though Maathai had repeatedly mentioned in the 1985 edition that the GBM's goal was to educate people on various elements of their interaction with the environment, she did not state consciousness raising itself as an explicit goal. In this expanded edition, it was the first element she described when she outlined the GBM's values and goals. She stated,

> Therefore, the overall goal of GBM in Phase I was to raise consciousness of community members to a level that would drive them to do what was right for the environment because their hearts had been touched and their minds convinced – popular opinion notwithstanding. (2004a, p. 33)

This sentiment reflects conscientization in the Freirean sense: The GBM's efforts were not about getting people to plant trees because they were told to, but because they valued it on their own accord. Freire (1970) argued that "human beings *are* because they *are* in a situation. And they *will be more* the more they not only critically reflect upon their existence but critically act upon it" (p. 109, emphasis in original). While the GBM certainly wanted women to engage in tree planting for environmental reasons, this section suggests that their motivation went beyond it: By engaging in the GBM, the women not only reflected upon their situations but actively worked to change it, and as such, the GBM facilitated empowerment.

Maathai echoed this idea later in the book when she discussed "lessons learned" (2004a, pp. 80-88). Of particular interest for this section is Lesson 2: "The messages must make sense to the participants" (p. 82). Here Maathai went into more detail on how the GBM engaged the women in a conscientization process:

> For instance one can ask community members to list the various ways that their families use/d the local biodiversity (e.g. as medicine, for construction, in traditional value- and spiritual-based ceremonies, as food and fodder). Such participatory discussions bring indigenous trees back into communities' daily lives and helps them to perceive the environment as a real and living part of their communal life. Yet another way is to discuss the possibility of attracting birds and other animals back to a given area, through reforestation, for the sake of current and future generations. Once such powerful but simple messages are understood, people become convinced and begin to take action. (pp. 82-83)

Not only did this form of discussion make an idea meaningful and relevant to people's lived experience, but asking the community members to generate the topics for discussion themselves, they are encouraged to reflect on their situation and the environments impact on them. As hooks (1994) has suggested, conscientization is "critical awareness and engagement" (p. 14). These seminars encourage both.

Although Maathai still focused much of her discussion on the involvement of women's groups as well as the impact of the GBM on the lives of women, improving women's self-image seems to have been less important here than it had been in the first edition of the manual. She still mentioned that some of the goals of the tree-planting program were "to generate income for rural women" (2004a, p. 37) and "to demonstrate the capacity of women in development" (p. 39). But the book focused predominantly on the positive impact that raising consciousness about people's relationship with the environment had on women. For example, she recounted how women of one particular area told stories about their difficult past, when they had to walk many miles to fetch wood and water. Maathai stated,

> [T]oday, however, they proudly tell how they can quickly obtain sufficient supplies of wood fuel at no cost since it is now available on their farms…. The men are grateful and full of praise for the women because of the wonderful work that they have done for the community. (p. 24-25)

Though the image of women had been elevated, this change was presented as a direct consequence of the tree-planting activities. It also was secondary to the positive impact of trees on the women's everyday lives.

Maathai did, however, add other areas of conscientization that she had not previously discussed: She emphasized the importance of (1) remembering traditional knowledge and indigenous food crops, and (2) civic education.

One of the goals the GBM had added since 1985 was to remind people of the validity of their conventional knowledge and traditions. While this idea at first seems to be in conflict with improving the image of women, Maathai claimed that

> it was necessary because the cultural values and systems of indigenous Kenyans were eroded, trivialized and deliberately destroyed in the process of colonization….
> The restoration of positive spiritual and cultural values is important since these contribute towards restoration of individual self-confidence, empowerment and identity. This restoration is also important in the protection of indigenous biological diversity, knowledge, practices and wisdom. (2004a, p. 48)

In terms of critical pedagogy, this element is of particular importance because conscientization does not mean forcing the oppressor's worldview onto the oppressed. Part of the problem the GBM tries to overcome by reminding people of their conventional wisdom and the importance of indigenous food crops is the oppression the African continent experienced at the hands of colonial forces. If Kenyans are to be empowered, then re-instilling pride about their culture is of the utmost importance. Freire (1970) argued,

> It is not our role to speak to the people about our own view of the world, nor to attempt to impose that view on them… We must realize that their view of the world, manifested variously in their action, reflects their *situation* in the world. (p. 96, emphasis in original)

This situation includes the loss of identity and with that, Maathai argued, the loss of appreciating traditional farming methods and indigenous food crops. Both, Maathai contended, are crucially important if people want to reverse the disastrous effects of deforestation and desertification. Increasing people's awareness, then, is necessary.

Another element Maathai added in the 2004 edition of the manual was civic education. Although the GBM had always relied on education and training seminars, much of these had originally focused on increasing awareness about people's relationship with the environment as well as teaching proper tree-planting methods. As the organization became involved in political and environmental advocacy over the years, the "GBM established a pilot civic education project" (2004a, p. 47). The topics included governance, culture and spirituality, Africa's development crisis, and human and environmental rights. The classes were a logical expansion of the original seminars because they "focused on the linkages between poor governance and environmental degradation" (p. 47), "poverty, unemployment, population pressures and environmental degradation" (p. 48), and "economic justice" (p. 49). As both Freire (1970) and Maathai (2004a) claimed, once people start to make connections and realize their potential, they become active. Providing civic education is just another step in the conscientization process.

Praxis. Concurrently with presenting additional detail on the conscientization efforts, Maathai also described in greater depth the GBM's practices that had led to praxis. Most notable were her discussions of the rationale for choosing tree planting as well as something she called "foresters without diplomas."

While tree planting is a practical solution, that in and of itself does not make it praxis in the Freirean sense. Praxis requires educational practices that make knowledge meaningful for the student's lived experience. Maathai specifically referred to this notion in her discussion of the first lesson learned: "Community development initiatives should address community-felt needs" (2004a, p. 80). She reiterated the idea that tree planting became the GBM's primary activity because it addressed the communities' immediate needs for food, water, shelter, and firewood. At the same time, she explained that the GBM "uses tree planting as an entry-point into communities since the trees meet many felt needs of rural communities" (p. 80). While she had hinted at the importance of making tree planting meaningful in the previous edition of the manual, she was more explicit in this edition.

Furthermore, tree planting as an activity can be meaningful and constitute praxis only if it is fully understood by the entire community. In this case, this refers to literal understanding based on language or translation issues. The Nation State of Kenya comprises 42 different ethnic groups in addition to significant minorities of Arab, South Asian, and European descent (Chege, 2008). Although English and Kiswahili are the official languages, many of the illiterate or semi-literate women the GBM works with are more fluent in their native tongue (Maathai, 2004a). This adds a significant challenge because a lack of understanding also means that any information shared or discussed becomes void and meaningless. Maathai explained that "to increase efficiency at the local level, the ten-step procedure was translated into local languages" (p. 41). In addition, seminars are held in "the language most stakeholders understand" (p. 96). In terms of praxis, this effort to overcome the linguistic challenge indicates the GBM's desire to engage its members on their level, demonstrating critical awareness of the women's lived situation.

Maathai expanded on her discussion of the establishment of tree nurseries by women's groups. As the previous section analyzing the 1985 manual suggests, the establishment of the nurseries themselves constitutes praxis because it required the women to rely on their own knowledge. But in that edition, Maathai had not elaborated on why the women had to rely on their own knowledge. In this expanded edition, Maathai described what she called "foresters without diplomas." When the GBM first decided to establish its own nurseries, it "organized seminars to which government foresters were invited to teach the basics of tree nursery management to the women" (2004a, p. 27). Unfortunately, these seminars proved less than helpful because the foresters insisted on using technical terms and discussing specialized tools the women could never afford. Then, as Maathai described it, "the women decided to do away with the professional approach to forestry and instead.... use their traditional skills, wisdom, and plain common – and perhaps women – sense" (p. 27). When the women relied on their own knowledge, tree planting became just another agricultural activity, and as farmers they had plenty of experience looking for seeds, and growing and transplanting seedlings. Realizing the importance of their own abilities allowed the women to shed another layer of oppression, adding to the process of conscientization through praxis.

Dialogue. Again, the last element Freire (1970) emphasizes is dialogue. Through the details given in this expanded edition of the manual, Maathai provided much more insight into how dialogue contributes to conscientization as well as praxis.

Maathai emphasized the importance of making the message meaningful to the participants. In that section, she specifically described topics the GBM has used to increase people's awareness of the importance of trees to their lives: for example, "one can ask community members to list the various ways that the families use/d the local biodiversity" (2004a, p. 82). It is this form of dialogue that leads members to critically engage with their situation. Instead of providing the answer, the GBM encourages its members to scrutinize their environment and living circumstances. Freire (1970) asserted that true dialogue not only invites critical thinking but depends on it. Thus, this form of dialogue leads to conscientization and with that increases the chances for education as libratory practice.

The same strategies were used by the GBM during its civic education training seminars. Instead of imposing others' perceptions of the world, the subjects of discussion evolved around those relevant to participants. Maathai contended "the seminars, which were interactive in nature, ensured that the experiences and concerns of the participants formed the basis for discussion and recommendations" (2004a, p. 47). This provided participants with a safe space to air their grievances and a support structure that could help them articulate possible solutions. In terms of critical pedagogy, providing this safe, dialogic place is vital because "those who have been denied their primordial right to speak their word must first reclaim this right and prevent the continuation of this dehumanizing aggression" (Freire, 1970, p. 88). Thus, the GBM's civic education seminars fall within the realm of critical pedagogy because they encouraged true dialogue, allowing participants to reclaim their place in the world.

Maathai discussed how the dialogue encouraged by the GBM's practices culminated in praxis. While the seminars provided a formalized space for dialogue, they were not the only instances in which dialogue occurred. In fact, often the co-creation of knowledge encouraged by true dialogue happens on a community level. For example, one of the consequences of eliminating the training the women were receiving from governmental foresters when they first established tree nurseries was the need to translate existing knowledge about farming into survival strategies for tree seedlings. As Maathai explained, "[T]he women quickly became very innovative and used techniques that would have been completely unacceptable to professional foresters…. Twenty years down the road, the women have gained many skills and techniques that they continue to share among themselves" (2004a, p. 28). Knowledge that was gained by engaging in dialogue is still shared and adapted as new information is processed. Encouraging the women to rely on their common, or women, sense started a process that allowed the women to discuss possible solutions and share experiences.

But dialogue also leads to praxis on an organizational level. One of the principles of critical pedagogy is the idea that both the oppressed and the oppressors, students and teachers, learn from the each other and are, thus, liberated. Tree planting and civic education seminars would not have constituted true dialogue unless the participants had learned something new. In her 1985 manual, Maathai had already mentioned that as a consequence of the seminars, the GBM had incorporated topics such as how to keep livestock away from seedlings and how to play soccer without destroying trees. Issues such as these arise during everyday life, and they need to be accepted and acknowledged for critical pedagogy to occur. In the 2004 edition of the manual, Maathai explained that GBM staff also took away much greater lessons from the first several years. She suggested that listening to the women made it clear that the GBM needed to expand beyond tree planting: "With time, other projects were initiated to address needs either arising out of the tree-planting campaign or in response to new environmental and/or developmental needs" (2004a, p. 34). The new projects included civic education and advocacy. Both of these now-important elements of the GBM arose as a consequence of people's shifted perception of the work around them.

Summary of Findings

From the analysis of the first (1985) manual, it seems clear that the GBM engages in strategies that fall within the realm of critical pedagogy: It engages in all of the tenets outlined by Freire throughout most of its processes. Through educational efforts, the GBM encourages conscientization among its members, with regard to both their relationship with the environment and the self-image of women. Many of the activities and efforts rely on praxis immediately relevant to members' lived experience, while organizational procedures, seminars, and community involvement invite continuous dialogue. Although this analysis has examined each of the tenets separately, the repeated references to the GBM's activities of tree-planting ceremonies and education suggest that all of the elements work in conjunction, another indicator that the GBM's practices are critical pedagogical.

Yet, despite the fact that this analysis seems to suggest that the GBM uses a critical pedagogical approach, the insights gained from the first manual are limited. Because the booklet summarized the GBM's historical development and its procedures and objectives on a limited number of pages, and spent 24 of its 77 pages on the forms used, much detailed description – for example, of how conscientization was achieved in terms of people's relations to the environment – was lacking.

The analysis of the 2004 manual supports the previous results of the 1985 manual: The GBM engages in practices that are designed to propagate critical pedagogy. Analysis of the newer edition supports the notion that although the GBM has been engaged in critical pedagogical practices from the start, its strategies have expanded and evolved over time. Maathai still discussed many of the same practices as she had in the 1985 version, but elaborated on each of them in much more detail. This provides a better understanding of how the practices work to empower people. By describing the particular strategies used to engage participants in dialogue, Maathai provided insight into the conscientization processes. She also emphasized the importance of consciousness raising, something that had only been alluded to in the previous edition. Furthermore, Maathai was much more explicit with regard to making knowledge meaningful and relevant to the participants' lived experience as well as making it understandable so that praxis could in fact occur. Lastly, in describing the dialogue GBM staff and members engage in, Maathai drew clear connections between all of the practices of the GBM. It becomes apparent that although conscientization, praxis, and dialogue have been treated separately in the analysis, they all are interrelated.

The GBM not only incorporates all of the elements listed by Freire, but in the 2004 manual all of the elements worked jointly to demystify the hegemonic structures that continued to oppress Kenyans. While Freire (1970) focused his educational efforts on literacy programs for Brazilian peasants, Maathai and the GBM identified environmental destruction as a systemic problem resulting in the oppression of Kenyans in general and women in particular. By drawing the women's attention to the problems resulting from deforestation, the GBM engages in conscientization. The goal is for women to rethink not only their relationship to their literal environment, but also their overall standing in society. Planting trees results in the realization of their own potential to overcome their situation, not only in terms of providing for their families, but also with regard to improving their societal standing. While the practical activity of tree planting is of the utmost importance to the GBM, at the heart of its approach is increasing people's awareness to their possibilities. As Maathai (2004a) stated, "the GBM views individual and communal empowerment as an important aspect of development because it is, in many cases, the mind-set from which people begin to realize their potential" (p. 38). This attitude, paired with practices that reflect the tenets outlined by Freire, suggests that the GBM is engaged in critical pedagogy. Further detail on the significance of this finding is provided in Chapter 5.

CHAPTER 5 – DISCUSSION

Using the results of Chapters 3 and 4, the discussion in this chapter seeks to answer the research questions asked in Chapter 1. Each question is considered in turn. Following the answer to RQ5 is a discussion of the limitations as well as recommendations for future studies.

Research Questions

RQ 1: How did the Green Belt Movement's frame(s) develop over the course of its existence? In Chapter 3, the frames established by Maathai in each of the three speeches under investigation were analyzed by examining the topic selections as well as the salient themes used to establish supporting claims. This current section provides conclusions about the development of the frames over the course of the time period selected and, thus, answers RQ1. Taken together, the three speeches build on each other to lead Maathai to frame the GBM's holistic approach as the solution to environmental and social problems. This conclusion can be derived from tracing the development of recurring frames and themes over the course of the 20 years.

Kuypers (2009) suggested that rhetoricians can glean the most useful information about a rhetor's framing efforts by comparing how themes and frames evolve over the course of time. The idea of interconnectedness presents just such an opportunity. Because this idea was not just present but had prominent placement in each of the speeches, it stands to reason that Maathai's framing of the GBM and its approach centered on the notion of interconnectedness. What changed over the course of the three rhetorical acts were the elements used to establish interconnectedness, as well as its status compared with other common elements. This change in status suggests frame re-articulation efforts on Maathai's part.

Maathai's Right Livelihood Awards speech revolved around the idea that all actions, good and bad, are inextricably connected to each other, and that all people are connected to each other and, thus, need to get involved. Because the speech represented one of the first opportunities to introduce the GBM to a largely international audience, it was crucial for Maathai to focus the speech on establishing the GBM's foundational principle. Therefore, she framed the speech in term of interconnectedness so that the audience would understand that interconnectedness undergirds every aspect of the movement's activities. By explaining in depth the cascading effects of positive and negative actions, Maathai provided the audience with clear evidence, repeatedly using verbs such as "precipitates" and "necessitates," which suggested the connection of one action with another. Maathai wanted the audience not only to see interconnectedness as the foundational principle behind the GBM, but to appreciate it as crucial to finding a lasting, effective solution. To do this, she established the need for a global/local approach. Again, if the audience could be made to realize that their actions – good or bad – had consequence a world away, then perhaps they would be more willing to engage in practices that would limit lasting damage.

While interconnectedness was presented as the foundational principle of the GBM in the first speech under investigation, its status was already slightly reduced during Maathai's Edinburgh Gold Medal Award acceptance address in 1993. While the speech still contained the same important elements of interconnectedness, the GBM's work, and a global/local approach, their arrangement suggests that Maathai was beginning to articulate an even larger frame. First, interconnectedness took a back seat to the frame of inequality as the root cause for the problems facing the planet; and second, while the solution was still framed in terms of interconnectedness, much of the last section of the speech set up the GBM's approach as a practical example for the solution. Previously, the entire Right Livelihood Awards speech had been about interconnectedness, and the GBM's work had merely provided evidence that demonstrated social connections. This time, Maathai focused on establishing the inequality frame and setting up the GBM's holistic approach as the solution. If the audience could accept inequality as the cause, then they might be willing to accept the GBM's use of interconnectedness as part of the solution.

By providing the audience with a clear cause for global problems, Maathai asked them to take the importance of interconnectedness to heart by recognizing its utility to their own lives. Yet, achieving acceptance of interconnectedness as the solution to global problems posed one of Maathai's most substantial rhetorical challenges. Framing the world in terms of interconnectedness is not necessarily an idea that many people are willing or able to take for granted. Entman (1993) and Kuypers (2006) argued that framing not only provides the audience with a different perspective of the event, subject, or item; but if the rhetor is successful, the audience accepts this new angle as theirs, abandoning other ideas. In part, Maathai's frustration might have been the result of continuously observing failed global efforts to overcome the power and development imbalances on the planet. At the same time, although the GBM was successful in Kenya during the 1980s and 1990s, its efforts to expand the approach into other African nations was often hindered (Maathai, 2004a), reiterating the challenge of reframing people's perceptions.

The most significant framing development occurred in Maathai's Nobel Peace Prize lecture in 2004. Whereas interconnectedness had been framed as the solution in the previous two speeches, it was relegated to a theme in this rhetorical act. At the same time, the GBM's holistic approach had graduated from a supporting theme to the dominant frame. This development suggests Maathai believed that the audience had come to accept the importance of interconnectedness, and that the time had come to amend the frame. Benford and Snow (2000) explained that a new frame tends to build on existing attitudes and beliefs, and does not necessarily introduce radically new ideas. Instead, rhetors take what is known, rearrange it, and arrive at a slightly different conclusion. Maathai still used the same elements to describe the GBM that she had used for the past 20 years, but she recognized that if the importance of interconnectedness had gained acceptance, then her argument could evolve to the next level. This next level was the idea that the GBM's holistic approach could provide the solution to the problems the world was facing. This strategy falls into line with what Snow et al. (1986) called *frame extension*: a strategy that provides movements with the opportunity to broaden the primary frame by including points of view that are congruent but not officially included. The notion of holism is congruent with the idea of interconnectedness, but so far had not been at the forefront of Maathai's previous framing efforts. Because it fit with the original concepts, it was less likely that the audience would reject the new frame.

One of the reasons why Maathai might have felt comfortable with this new approach lies in the significance of receiving the Nobel Peace Prize itself. During her address, Maathai repeatedly referred to "the visionary action" (2004b, para. 8) of the Nobel Committee, because by awarding the prize to her it had "placed the critical issue of environment and its linkages to democracy and peace before the world" (para. 8). In its press release ("Press release," 2004, October 8), the Peace Prize Committee argued that it had chosen Maathai and the Green Belt Movement because of the

> contribution to sustainable development, democracy and peace. Peace on Earth depends on our ability to secure our living environment, Maathai stands at the front of the fight to promote ecologically viable social, economic and cultural development in Kenya and Africa. She has taken a holistic approach to sustainable development that embraces democracy, human rights and women's rights. She thinks globally and acts locally.

If one of the most prestigious awards in the world considered the GBM's holistic approach to be not just successful but worthy of being highlighted, then setting up this approach as *the* solution seemed to be at least worth considering. Furthermore, the large audience Maathai was addressing during the lecture provided the perfect forum for extending the frame: a large, most likely supportive audience increased the chances that this new frame would be accepted and, as a consequence, spread.

The frame that did not explicitly appear in the Nobel Lecture was the inequality frame Maathai had set up during the second speech under investigation. Considering that Maathai had already been working toward this frame in the Right Livelihood Awards acceptance speech – albeit in its infancy – and the significance of the frame during the Edinburgh address, the lack of this frame in the Nobel Lecture seems somewhat curious. At the same time, though, it is hardly surprising. Because of its prestige, the Nobel Peace Prize can be considered a culminating moment for a social activist. While there are certainly varying degrees of success for each laureate, the celebration of the GBM's success as well as the progress in Kenya's peace process all lent themselves to using a more hopeful, positive tenor in this speech.

Although not a separate frame or even an explicit subject of discussion, the theme of inequality was nonetheless present in the speech. In describing the GBM's work, Maathai touched on the challenges women are facing every day, implying that many of those challenges are the result of inequality. She did not, however, feel the need to be more overt. In addition to choosing a more hopeful tone for the speech, she might have felt that inequality was an accepted frame that did not need further development. Benford and Snow (2000) suggested that because social movements fight against the existing structure, they more often than not rely on injustice frames to argue for change. Oftentimes, this injustice frame identifies "the source(s) of causality, blame, and/or culpable agents" (p. 616). Inequality represents one of those injustice frames because it identifies the causes. As such, it was likely to be more generally accepted, and further development would have detracted from establishing the new holistic frame.

The analysis of Maathai's speeches clearly shows that while the frame employed to describe the GBM and its approach has changed over the course of time, the primary rhetorical elements have stayed consistent. This supports the idea presented by Benford and Snow (2000) that framing is most effective if existing attitudes and beliefs are not abolished but reinterpreted. It also suggests that the GBM has stayed true to the same principles since its early days, arriving at the conclusion that a holistic treatment is needed to overcome the challenges we are facing.

RQ 2: Do the GBM's practices follow critical pedagogical principles? In Chapter 4, the GBM's educational practices were analyzed for conscientization, praxis, and dialogue by exploring two manuals written by Maathai. Based on the findings of the analysis, it seems clear that the GBM engages in critical pedagogical practices. As established previously, according to Freire, critical pedagogy is derived from radical and progressive educational movements aspiring to link democratic principles with transformative action (Darder et al., 2009). This means that the empowerment of students is the result of demystifying hegemonic, oppressive structures by using a practical educational approach (McLaren, 2003). For Freire (1970), that meant focusing teaching principles on the students' lived experience by engaging them in dialogue and raising their consciousness to their life situation. Focusing on raising people's awareness of the link between environmental destruction and their own situation, the GBM has used critical pedagogical principles successfully to demystify hegemonic structures.

While both manuals allow insight into the educational practices of the GBM, it was the 2004 edition that provided greater detail and most clearly demonstrated the GBM's commitment to the tenets of critical pedagogy. As mentioned previously, the 1985 manual seemed to be the result of the GBM's desire to expand its approach into other primarily African nations. While Maathai described in depth the structure of the GBM and the tree-planting activity, she omitted some of the details that showed how critical pedagogy was achieved. For example, she repeatedly mentioned the need for environmental education but did not explain the strategies used by the GBM to encourage consciousness raising.

Another challenge of the first manual was the inclusion of the forms and application letters used by the GBM. While these appendices demonstrated precisely how the GBM was structured as well as what procedures they relied on, Maathai failed to mention the implications of having to fill out such forms. Many of the rural women the GBM encounters today are only semi-literate. It stands to reason that when the GBM began its work in 1977, even fewer of them could read and write. That they were asked to fill out forms that were several pages long, written in one of the official languages of Kenya, reasserted oppressive structures the GBM desired to overcome. It was the 2004 manual that provided the detail needed to better comprehend how they dealt with this challenge. The analysis touched on the fact that the GBM has translated much of the ten-step program into other languages. By providing material in people's native language, the GBM has reduced the oppression stemming from having to use official, non-native languages. Furthermore, Maathai mentioned that most communities have literate members who are usually willing to fill out the forms even if they themselves are not members of the GBM. Where that is not the case, GBM staff helps groups to fill out any of the forms needed. While that still requires reliance on others and does not eliminate the oppression resulting from a literacy requirement, the GBM demonstrates its awareness of the problem by ensuring the groups can participate in their program regardless of their literacy.

One of the most significant differences between the GBM of 1985 and that of 2004 is its involvement in civic education and advocacy. While the GBM was involved in consciousness raising and education from the start, much of it focused primarily on desertification and reforestation. Civic education and advocacy were added to their core objectives in response to the community's increased interest in the connections between governance and environmental degradation. The GBM's willingness to evolve with the needs of the community suggests dedication to critical pedagogical practices. Freire (1970) argued,

> The pedagogy of the oppressed... has two distinct stages: In the first, the oppressed unveil the world of oppression and through the praxis commit themselves to its transformation. In the second stage, in which the reality of oppression has already been transformed, this pedagogy ceases to belong to the oppressed and becomes a pedagogy of all people in the process of permanent liberation. In both stages, it is always through action in depth that the culture of domination is culturally confronted. (p. 54)

The addition of civic education and advocacy to the GBM's principal objectives suggests that the GBM and its members were moving from the first stage of critical pedagogy to the second. The 1985 manual focused almost exclusively on raising people's awareness about the importance of the environment to their living conditions. This reflects what Freire contended about the first stage of critical pedagogy: Raising people's awareness of the systemic oppression facilitated by environmental destruction equaled unveiling the world of oppression, while the tree planting functioned as the praxis that allowed individuals to commit to transformation.

As the analysis suggests, the GBM added civic education and advocacy because the people's needs changed over the course of time. While at first they were primarily interested in fulfilling the basic needs of water, firewood, fuel and shelter, they began to realize their potential and with that desired to effect more significant change. At this stage, Freire suggested, the pedagogy moves from just one group to all people. While in the case of the GBM this second stage may not have been as far-reaching or revolutionary as Freire and other critical pedagogical scholars would have desired, the GBM's progress suggests significant change. Maathai (2004a) posited, "initially, it was only GBM and a few other concerned parties that would organize advocacy campaigns... Currently, however, members of the public have become increasingly involved in advocacy because they have seen its positive impact" (p. 51). Additionally, as Maathai explained in the preface to the 2004 edition, members of the GBM gradually moved from environmental campaigns to become politically engaged to advance democracy in Kenya. This implies another shift in focus, broadening transformative action yet again.

RQ 3: How do the Green Belt Movement's educational practices reflect the frame(s) established? Using the results of Chapters 3 and 4, this question seeks to understand whether the GBM's educational practices are in accordance with the holistic frame established by Maathai. Although the frame put forth by Maathai has changed slightly over the years from an interconnectedness frame to a holistic frame, many of the themes and topics of discussion have remained the same. Since the beginning, Maathai's framing efforts have included interconnectedness, awareness raising, and using a global/local approach. This section explores whether these framing elements are present in the educational practices described in the manuals.

One of Maathai's most important strategies for setting up interconnectedness initially was to draw her audience's attention to the impact of cascading effects. During the Right Livelihood Award acceptance speech in 1984, Maathai outlined both positive and negative cascading effects. These were also present in the both of the manuals. Maathai explained that tree planting became a meaningful activity for Kenyan women once they realized the interaction of deforestation, desertification, and their lack of water and food. At the same time, the women began to understand that they themselves could have a positive impact on their lives by engaging in the GBM's tree-planting program (Maathai, 1985, 2004a). Additionally, the GBM has placed special emphasis on engaging children in its program. Maathai (1985) suggested "time was ripe… to encourage school children and educate them on the need to conserve the environment" (p. 22). By understanding the connections between the environment and their own lives, these children would likely be better stewards of that environment in the future. More importantly, the GBM realized that some of the children would take their new-found knowledge home with them and share it with their families (Maathai, 2004a). This reinforces the notion that all actions are inextricably connected.

As Chapter 3 suggests, raising awareness is a cornerstone of Maathai's framing efforts: each speech included topics and themes suggesting not only that Kenyans needed to become more aware of their surroundings and living circumstances, but that it was equally important for the international community to interrogate their lives. During the speeches, awareness raising was often combined with accepting personal responsibility and a call for a shift in thinking. While the speeches included discussion of both Kenyans and the international audience, the manuals focused almost exclusively on GBM members. The analysis in Chapter 4 indicated that the GBM has engaged in conscientization efforts since the very beginning. The core of these conscientization efforts was the tree-planting program. Through it, women not only increased their awareness of interconnectedness, but found that they had the potential to take charge of their own lives. This realization was the first step in empowerment and led the GBM to add additional training programs designed to increase people's awareness about democracy and advocacy (Maathai, 2004a). As such, the manuals demonstrate how the GBM has incorporated awareness raising into its educational practices.

According to Maathai, the GBM's holistic approach will only be effective if it is accepted on local and global levels. Hence, Maathai spent much time in the speeches outlining the responsibilities of Kenyans and the audience. The GBM's emphasis on community involvement reflects Maathai's notion of the local level. Tree planting is an inherently local activity: The trees will thrive only if those who planted them take care of them. Most immediately, this idea refers to the members of the woman's group who have become members of the GBM. On a broader and, to the degree of their surroundings, more global level, it suggests that the entire community needs to become engaged in tree planting. The manuals, thus, indicate that the GBM's educational practices focused on using a local approach.

Additionally, the manuals reiterated the importance of a global component. In the preface to each manual, Maathai (1985, 2004a) stated that the book was written to share the approach and to answer questions that others might have about the GBM. While the manuals did not advance strategies on how to adapt the GBM's educational practices to other national contexts, they had the potential to function as a tool to replicate the GBM's approach. Their existence, thus, implies that Maathai and the GBM recognized the need for such a tool. Although the 1985 manual was published only in English, the 2004 expanded edition has been translated into French, Japanese, German, and Korean in an attempt to reach a global audience ("Green Belt Movement: Books," n.d.).

Although this discussion does not include every theme or topic mentioned by Maathai in the three speeches previously analyzed, it suggests that the GBM's educational practices reflect Maathai's frame. Each of the ideas explored is an element used by Maathai to advance the frame that environmental and social change requires a holistic approach. While the above answer examined each element individually, it becomes apparent that it is the combination of all of the elements and their consistent use that have made the GBM successful.

RQ 4: What insights can be gained from the exploration of the GBM's frame and educational efforts with regard to their applicability beyond Kenya? As the description of RQ4 in Chapter 1 implies, one of the challenges framing scholars of social movements struggle with is the transferability of frames beyond the individual movement efforts. Benford and Snow (2000) explained that "just because a particular [social movement organization] develops a primary frame that contributes to successful mobilization does not mean that that frame would have similar utility for other movements or SMOs" (p. 619). Although the preceding analysis does not verify that the GBM's frame has been effectively transferred to other movements, it does indicate that the GBM's frame as established by Maathai has the potential for transferability.

During the speeches, Maathai repeatedly appealed to the international community to accept the need for a holistic solution to the problems of poverty, development, environmental destruction, etc. The fact that she was awarded the Nobel Peace Prize based on the GBM's approach suggests increased interest in and acknowledgement of the GBM's founding principles (Maliti, 2004). As such, it stands to the reason that the frame suggested by Maathai and the GBM is seen as having potential benefits for other social movement efforts. This notion is supported by Dove (2008), who argued that recognizing the linkages between peace, environment, and democracy has not only led to "a new field of study that has emerged in the past dozen years called 'environmental security,'" but "reflects the idea of 'new environmentalism'" in which people have expanded their vision to include "educat[ing] themselves about their relationship of tree-planting to larger, more abstract issue and institutions of democracy, justice, and peace" (p. 7). As Maathai explained in the manuals, the GBM started the tree-planting program to address many of the immediate concerns of the people and mitigate ongoing environmental destruction.

This idea of focusing on the immediate concerns of the people combined with environmental protection and advocacy stands at the forefront of the GBM's approach and should be transferable to other movements owing to the concepts' broader applicability.

Given that social actors have almost unlimited access to materials of and reports about other social movement organizations, it should come as no surprise that today's activists carefully plan their actions by borrowing rhetorical strategies and tactics from others. One of the most recent examples was the protests in Egypt that led to the resignation of Hosni Mubarak in February 2011. Stojanovic and Gec (2011, February 22) reported that Egyptian activists had consulted with the Serbian pro-democracy organizations that successfully ousted Slobodan Milosevic to better understand the principles that aided in the peaceful uprising in Serbia. The group had also successfully trained activists in Georgia and Ukraine during the revolutions in 2004 (Stojanovic & Gec, 2011, February 22). While the analysis of the GBM does not provide specific examples of the successful implementation of the principles in other nations, spreading their approach is one of the organization's declared objectives. Not only does the GBM hold training seminars for social movement leaders of other African nations, but the fact that the manual has been translated into other languages suggests their desire to facilitate access (Maathai, 2004a).

Additionally, the GBM's holistic approach has applicability beyond the context of social movements. Citing the GBM as an example for successful implementation of a holistic approach, Ketola (2008) argued that such an approach should be taken into consideration for holistic corporate responsibility models. She posited that "an *ideal* company maximizes its economic, social and ecological responsibilities" and suggested that although implementing such an approach is easier for cooperatives than for corporations, striving for holistic corporate responsibility increases the company's lifespan.

Furthermore, the GBM's approach has potential as a strategic approach for coalition building within social movements. Meyer (2007) contended that labeling any effort as a "social movement" out of grammatical convenience "distorts the reality of a social movement, reifying boundaries and movements that are actually much sloppier affairs" (p. 74) because the term combines a group of organizations that fight for a similar goal. This grouping oftentimes occurs as a result of coalition building. Meyer (2007) further explained that joining such cooperation efforts can be beneficial to organizations because it may strengthen the overall cause, but simultaneously bears the risk of decreasing the visibility and distinctiveness of individual organizations. As previously stated, the GBM defies simple definition because of its engagement in a variety of causes. In fact, one of the reasons why the Nobel Committee awarded Maathai the Peace Prize was the GBM's involvement in environmentalism, human rights, women's rights, sustainable development, democracy efforts, and education. By suggesting that social change can be realized only if we take a holistic approach to the problems, causes, and solutions, the GBM minimizes the differences between the various efforts, increasing the potential for coalition building.

The GBM's educational approach has applicability beyond Kenya as well. Comparing Maathai's 1985 manual with the 2004 edition demonstrates that the GBM's educational practices have expanded over the course of the organization's existence, advancing into the second stage of critical pedagogy. As previously mentioned, during the second stage, critical pedagogy moves from just one group to all people, increasing the potential for empowerment and the reduction of oppression. While this claim cannot be substantiated until other social movement organizations' educational practices have been examined, it is reasonable to argue that the incorporation of critical pedagogical principles into a movement organization's approach is vital. This notion supports Holst's (2002) claim that education is principal in advancing social movement efforts, as well as Foley's (1999) contention that movements need to pay much closer attention to education and learning in their efforts.

Lastly, the synthesis of the GBM's framing efforts with its pedagogical approach has had a positive impact on the organization's survival and success over the years. As Kenyans became increasingly aware of their surroundings and their political potential, the GBM expanded its educational efforts into other subject areas, such as civic education and advocacy. Reacting to the changing demands by expanding their core objectives without abandoning their founding principles and programs has helped the GBM succeed for the past 30 years. Stewart et al. (2007) argued that one of the greatest challenges social movements face is time and, with it, the changing nature of demands. In order to survive in the long term, movements need to establish strategies that allow them to react to these changes without betraying their original cause. The GBM's emphasis on interconnectedness and its holistic approach have allowed the organization to incorporate other efforts much more easily. In fact, adding civic education and advocacy seem almost like a natural extension. While it cannot be ascertained that the exact same approach would work for other organizations, the GBM's consistent holistic approach allows for greater adaptability and, with that, a greater chance for survival.

RQ 5: What are the consequences of the GBM's framing and educational practices for social movement theory? One of the greatest challenges social movement theory has faced is the question of defining what constitutes a social movement. Because of differences in opinion among scholars, social movements have been divided into old (OSM) and new (NSM).

Following the characteristics outlined by Stewart et al. (2007), OSMs are said to be more structured, to operate outside of and in opposition to the establishment, and to focus on universal demands. NSMs, on the other hand, if they are defined at all, are said to be decentralized, to be able to work from within, and to pursue particularized demands, if they are defined at all (Huesca, 2001). Within the Green Belt Movement, elements from both schools of thought can be found. This makes placing the GBM within either paradigm rather challenging, but at the same time it provides a unique opportunity to move beyond the debate and instead focus on advancing our understanding of social movements as well as expanding our approaches rather than restricting them.

It is the GBM's involvement in a variety of causes that makes it difficult for a scholar to define it as either OSM or NSM. By approaching social change from a holistic perspective, the GBM touches on both elements of the universal versus particularized demands debate at the same time. Certainly, fighting for food, water, and survival is a very obvious economic and materialistic issue. The primary reason Maathai created the GBM was to provide women with a means to care for their families (Maathai, 2007). Because of deforestation, the land was parched and women were lacking the resources to improve or at least maintain their lifestyle. Yet, at the same time, addressing this universal concern led the GBM to include particularized demands of democracy and identity politics. From its inception, the GBM has emphasized the importance of women's rights by providing women with practical solutions that allowed them to realize their own potential. Since then, the GBM has added civic education and advocacy to advance human rights and democracy efforts. As Maathai stated, "[O]nce you start making these linkages, you can no longer do just tree-planting. When you start working with the environment, the whole arena comes: human rights, women's right, environment rights, children's rights...everybody's rights" (cited in French, 1992). Realizing the interconnectedness of women's rights, human rights, environmental rights, etc., is what leads to the necessity of approaching the struggle from a holistic point of view, interweaving the various issues to achieve broad social change. When women are taught how to provide for themselves, they learn to rethink themselves in terms of the larger society. They develop a greater understanding of place, of the politics around them, and of the patriarchal structures constricting them (Maathai, 2004b). The GBM's approach seems to suggest that instead of prioritizing one type of demand over another, both demands still go hand-in-hand for at least some social movement organizations. A refusal to examine how both influence the organization would lead to an incomplete picture. This supports Cloud's (2001) argument that ignoring universal demands, such as class and economic struggle, could prove detrimental to social movement scholarship.

Not only does this mean that social movement scholars need to be careful not to force a movement organization into a particular definition, but that they should also be willing to take a wide range of methodological approaches into consideration. Campbell (2006) suggested that the critic should find the most appropriate lens for the text in terms of context, culture, and language. This suggests that the choice of method should be determined by the questions that need to be answered about the artifact. One of the elements making the GBM intriguing for a social movement scholar is the fact that it is engaged in a variety of issues. How the GBM has justified the incorporation of various causes seemed to be one of the central questions that needed to be asked. Because framing gives insight into the ways that a rhetor wants the audience to understand an event, item, or subject, it proved to be the ideal method to answer that particular question. Yet, while framing answered the question of how Maathai wanted the international audience to understand the GBM's approach, it was not sufficient to explain how that approach was achieved in Kenya. In reading about the GBM, it quickly becomes clear that education plays a major part in their program. While critical pedagogy is not usually used as a lens for textual analysis, it provided a useful tool of analysis for understanding the GBM's educational efforts. Furthermore, to evaluate the educational practices of the GBM in terms of the organization's overall goal of empowerment, it was necessary to find a tool that would allow assessment. Because empowerment and the reduction of oppression are at the heart of critical pedagogy, this method was well suited to answer the question. While the combination of framing and critical pedagogy in the same study is unique, it proved to answer important questions about the organization. As such, one of the lessons for social movement theory that can be taken away from this study is the notion that methodological tools need to be derived from the questions asked about the artifact rather than from the scholar's desire to conduct a particular kind of study.

Lastly, while not directly derived from the analysis, one of the consequences of studying the GBM's framing and educational efforts should be greater interdisciplinarity and collaboration. The analysis of the GBM suggests that social movements are complex constructs; and although this particular study employed a rhetorical focus, it certainly does not touch on all of the elements of the GBM. To provide a more complete picture of this organization, it should be studied through a variety of lenses and by a variety of disciplines. Greater collaboration among disciplines would also reflect the GBM's holistic approach.

Limitations

One of the greatest textual limitations of this study concerns authorship of the material studied. As mentioned before, Wangari Maathai is not only the declared leader of the Green Belt Movement but has become its public face (Ndegwa, 1992). As such, all of the material under investigation has been either delivered or published by her. While choosing only material authored by Maathai could lend greater consistency to the study, being restricted to having to use material solely authored by her can pose the problem of providing one-sided insight. The Green Belt Movement's frame and its educational practices were first filtered through her voice. Consequently, the results of the study represent her perception of the GBM only.

Another textual limitation of this study pertains to the speeches used for the framing analysis. Although the three speeches used in this project were chosen with great care to ensure consistency in terms of type and time span, they certainly were chosen from among many given by Maathai. While Kuypers (2009) suggested that one of the great assets of framing as a methodological tool is its ability to assess the development of a frame over the course of a designated time span, he warned that insufficient text use can lead to meaningless results. At the same time, he warned that rhetorical scholars need to keep the amount of text manageable, both for themselves and for the audience reading the study. Unfortunately, there are no criteria as to what constitutes sufficient text use.

In the same vein, this study focuses the framing analysis on award acceptance speeches. This criterion was chosen for the sake of consistency. It needs to be mentioned, however, that Maathai's public appearances have not been limited to award ceremonies. She regularly attends and presents at conferences, is interviewed for television and radio programs, and presents public lectures at universities across the globe ("Green Belt Movement: Wangari Maathai," n.d.). All of these appearances can lend further insight into her framing efforts of the GBM.

There are also limitations with regard to the study of the GBM's educational practices. Because this study employs textual analysis as the principal mode of investigation, findings are restricted to the two texts chosen for this purpose. While the two manuals described the educational efforts of the GBM and their development in detail, they still constituted only descriptions by Maathai. This analysis does not provide any insight into the perception of the educational efforts by GBM members, nor does it provide the voices of GBM ground staff who have had to develop these educational practices. It is therefore possible that Maathai's descriptions diverged from the perceptions of members and staff.

This study relies on Freire's (1970) basic tenets of conscientization, praxis, and dialogue to evaluate whether the GBM's practices fall into the realm of critical pedagogy. Although this analysis demonstrates the possibilities of using the tenets proposed by Freire in a textual analysis, this methodological approach has its limitations. While the two manuals provided descriptions of the GBM's educational practices, they did not allow drawing conclusion about the reflective and dialogic processes the women and the GBM engaged in. Instead, this analysis had to rely on inferences about the development of the educational practices to illuminate the presence of critical pedagogy rather than the quality.

Additionally, Freire's method has been criticized as being too abstract and impractical, amongst other things. Consequently, there has been much effort to advance critical pedagogy beyond Freire. While not necessarily a limitation to the study, it needs to be pointed out that although there is much criticism about this method, one of the principal reasons for choosing Freire's tenets as the lens for the study are the similarities of circumstances under which Freire and Maathai operated during the inception of their respective programs. Both were highly educated and chose to empower mostly illiterate peasants and rural workers in their respective homelands. Both chose very practical programs emphasizing praxis instead of abstract empowerment. Lastly, both of them operated in countries overcoming the challenges of colonization and the resulting oppression.

A discussion of the study's limitations would be incomplete without a closer look at the subject position of the researcher herself. Academic research has often been criticized for misappropriation of meaning based on differences in cultural, ethnic, and other backgrounds between the researcher and his/her topic of interest. In the case of the present study, this difference could be interpreted as problematic. I am neither black nor African. I am white and I am German. I have no personal experience of Kenya and the challenges facing the women there. Yet, I contend that my cultural background (or lack thereof) does not pose a serious challenge to the above analysis for two major reasons.

First, all of the texts used in this study were written with an international, Western audience in mind. Maathai used her award acceptance speeches to introduce largely Western audiences made up of potential donors and supporters to introduce the GBM and its objectives, practices, and thought. Both manuals were written to share the approach with others, both in Africa and beyond. As Maathai already explained in her 1985 manual, GBM staff "are often approached for interviews, field trips, demonstrations, and pamphlets by individuals, researchers, planners, women leaders and all those concerned with rural development in general and community afforestation activities in particular" (preface). She spends much of her time teaching people across the world the GBM's approach ("Green Belt Movement: Wangari Maathai," n.d.). Considering that I am likely one of the intended audience members, there is less danger that my cultural background poses a challenge for the results of this study.

Second, and more importantly, one of the challenges of academics in general and social movement scholars in particular is the accusation that they apply Western standards to non-Western subjects of research, which then leads to the misappropriation of meaning. While that is certainly a concern, I would argue that the solution cannot be to abolish our methodologies or to abstain from rhetorically significant artifacts based on cultural difference. Instead, I believe we should still apply Western methodologies to non-Western movements. But instead of deriving implications about the effectiveness of the tactics and strategies chosen, we should re-investigate our analyses to see where our theories diverge from the movements' actual use of rhetoric. This approach will allow us to see which elements of our theories are context- and culture-specific, and which have more general applicability. Once we understand our theories better across cultural contexts, we can articulate new ones. For example, conducting a rhetorical analysis of the 2007 Burmese protests, Pinkerton (2009) discovered that owing to its cultural restrictiveness, social movement theory failed to properly address a variety of factors that played a crucial role in the protests, such as "non-oppositional rhetorical forms of communication, non-modernist conceptions of time, the possibility of collectivities functioning as distributed networks, and the complexity of identification with the movement" (p. 27). Uncovering these factors is the first step to formulating more comprehensive theories. This approach would also provide us with the opportunity to re-examine Western movements through a different lens to see whether our cultural context might have previously led us to misinterpret rhetorical strategies

Recommendations for Future Research

Based on the limitations listed above, future research should expand the voices analyzed beyond Maathai's. While the significant challenge of finding material not authored by her still exists, these research efforts could, for example, focus on documentaries of the GBM such as the one produced by Merton and Dater (2008). Including interview material of GBM members and staff in such documentaries can be useful to substantiate claims made in the present study. Another possibility to expand the voices heard would be to conduct place-based research in Kenya. This type of research could illuminate whether Maathai's public framing of the GBM reflects the perception of GBM members. It could elaborate on the educational practices of GBM and the members' perceptions of these efforts.

As mentioned earlier, Maathai produces rhetoric beyond award acceptance speeches and spends much of her time sitting for interviews, presenting lectures to universities, and attending conferences on development. All of these rhetorical situations provide her with additional opportunities to discuss the GBM and its approach. As such, future research should use this rich resource to broaden the scope of the present study to explore the consistency and development of her frames.

In her 2004 manual, Maathai described the process through which the GBM attempted to reach out to leaders in other nations so they could replicate the GBM's approach in their own countries. Research into these programs could prove useful on two levels. First, it might indicate how successful the GBM's training efforts are. One of the notions Maathai (2004a) discussed was the idea that for effective replication, hands-on training needs to be provided. While she went into detail about the challenges the GBM faced executing these training workshops, she did not provide any indication about the success of GBM projects in other African nations. Exploring these programs could provide an answer to that question. Second, examining GBM-like programs in other nations could further help us understand frame transferability. Benford and Snow (2000) argued that "just because a particular [social movement organization] develops a primary frame that contributes to successful mobilization does not mean that that frame would have similar utility for other movement or SMO's" (p. 618). Conducting framing analyses of GBM-like programs in other nations and comparing those studies with the GBM's framing efforts advanced in this study could provide more insight into the potential transferability of the GBM's holistic frame as advanced by Maathai.

Future research should include analysis of the GBM utilizing other social movement–related methodologies. The scope of this present study focuses on Maathai's framing efforts of the GBM as well as its educational practices. Other studies could include leadership studies of Maathai, the GBM's coalition-building efforts, or its mobilization strategies. While this particular study constitutes a rhetorical analysis, I believe that no single study can provide a complete picture of any social movement or social movement organization. In an attempt to illuminate the effectiveness and significance of a movement or movement organization, such as the GBM, we need to cast aside the desire to define social movement studies as purely rhetorical, sociological, or political, and value the insights and theories proposed by other fields. Campbell (2006) suggested that the critic should find the most appropriate lens for the text in terms of context, culture, and language. That requires the social movement scholar to take a wide range of perspectives into consideration before making the final choice. I would welcome reading scholarship on the GBM utilizing other methodologies, and I hope that my study contributes to the already existing understanding of the GBM.

APPENDIX A

THE RIGHT LIVELIHOOD AWARDS ACCEPTANCE SPEECH

Mr. Chairman, Ladies and Gentlemen,

A.

I have come to Stockholm to accept with the greatest humility and gratitude an award from the Right Livelihood Foundation. I have not only come on my own behalf but on behalf of the National Council of Women of Kenya (NCWK) especially the numerous women groups who produce the tree seedlings in the fifty odd tree nurseries, the thousands of school children who plant them and take care of them under the dedicated leadership of their teachers. I have come on behalf of the green belt staff who give a presence of the movement in remote places of our country, the individuals who have planted trees on their plots and on behalf of any person who has sponsored a tree in any of the green belts. I have also come on behalf of the donors who gave us funds to be able to translate our ideas into a programme. And so represented here are the thousands in my country and abroad who have shared our thoughts, our aspirations and our demonstrations.

We have been informing our people informally. We have been telling them that if they be found ignorant it must not be in the understanding of the Laws of Nature. We have been telling them that drought and famine need not be an annual event, some diseases need not be, malnutrition need not be and some deaths need not be. We have been telling them that with a little bit of help from outside and much will to use their resources on their part they can reverse the trend. They are listening, they are responding and they are struggling hard.

The Right Livelihood Foundation surprised us most pleasantly when it gave the 1984 award to use. We can never be afraid again because we know now that as we walk these paths we walk not alone. We are many and our number makes us strong.

B.

And now for the work we are involved in:

The green belt movement (GBM) is now a slogan which describes a broad-based grass-root tree planting activity currently taking place in Kenya. Since trees are planted in several rows around compounds or farm plots (shambas) the planting of trees appears to dress up the compounds in belts of green trees. In our adverse activities on the land e.g. in discriminate cutting down of trees, bush clearing, failure to stop soil erosion, overgrazing, over-population and overall general negligence towards our environment not only have we torn into rags the beautiful green dress of our mother-land but in some places we have stripped her naked. We have inflicted deep wounds on her and she is weak and unproductive, Yes, indeed, according to the prophet Isaiah, we have sinned against the Natural Laws (God, goodness, order of Nature) and we are being pushed. The Natural Laws are taking their natural course which for us means destruction and death. We must repent our sins (i.e. rectify our wrong doings) by dressing our mother our mother-land in her original beautiful and full green dress. In planting trees we are adorning our mother-land with belts, green belts.

When we have repented (i.e. rectified) our mother-land will be healed and we shall reap a bounteous harvest. And thus our committal, which we recite before planting trees:

"Being aware that Kenya is being threatened by the expansion of desert-like conditions, that desertification comes as a result of misuse of the land by indiscriminate cutting-down of trees, bush clearing and consequent soil erosion by the elements; and that these actions result in drought, malnutrition, famine and death, WE RESOLVE to save our land by averting this same desertification by tree planting wherever possible".

"In pronouncing these words, we each make a personal commitment to our country to save it from actions and elements which would deprive present and future generations from reaping the bounty which is the birthright and property of all".

C.

First why we do what we do:

1. The Kenya Government has a Ministry of Environment and Natural Resources and a Presidential Commission on Soil Conservation and re-afforestation. Both bodies are responsible for re-afforestation efforts in the whole country. But we know that few governments, and less so in the developing world, can afford the financial and man-power resources required to do what needs to be done.

It is necessary for private/voluntary, non-governmental organizations (NGOs) and individuals to be mobilized to provide at least the man-power needed in afforestation programmes.

2. As soon as we took trees to the people we realized that there was great demand for trees. People clamoured for the trees we issued at public meetings. This was a pleasant sight. Unfortunately, we also discovered that they did not appreciate the fact that trees like other crops, need to be planted properly, need after-care and have to be sheltered from livestock and human beings. It became obvious all to us that there was need to teach almost/ the people that they have to dig holes, apply manure, make sure that water is available and build shelter for protection.

3. The demand for trees necessitated the establishment of tree nurseries. In order to take more trees to a greater number of people we realized the need to train ordinary persons to

become seedling producers. Since we are a women's organization and many women are organized into groups we decided to make rural women groups our major target groups. We trained them on the basics of raising seedlings more or less like they can raise their cabbages and potatoes. These groups have now been joined by youth groups and clubs.

4. In order to promote the seedling production we decided to purchase seedlings at a minimal price of about US seven cents per seedling. This way not only do the groups gain new and useful knowledge but tree production becomes income generating.

5. Most people are crop-farmers and livestock-keepers. They cannot turn all their land into woodlots because they need it for crops. They are encouraged to practice agro forestry, farming methods our people used before the European methods of farming were introduced and erroneously considered superior. Now the scientists are recommending this agro forestry approach and unfortunately the current generation has to be taught to intercrop ……... all over again. This requires some knowledge on the trees and the role they play in the soil and in respect to other crops. Most indigenous trees for example, are of course better suited ecologically but many are slow growing and do not have much economic value in the current market. This puts them at disadvantage as farmers go for the exotic or imported trees which grow faster and have a well established market that is, at least to-day, when the trees grow in what is to them virgin land. Several hundreds years from now we may find that the exotic trees precipitated desertification and destruction of the varied life that flourish in tropical ecological systems. To discourage the planting of imported trees we pay less to seedling producers (mostly women) for them and more for the indigenous and fruit trees which are more appropriate for agro forestry.

6. The original major objective of the green belt movement was to help the needy urban poor of a certain area of Nairobi. In mind were the handicapped, school leavers and the very

poor. The best way to help them was create jobs. So we hired them as green belt rangers and nursery attendants.

Many of the green belt rangers and nursery attendants are illiterate and have no training in nursery or forestry techniques. We would provide them with basic training to be able to nurse the trees and assist the school children each of whom attends a few trees. By employing such persons we were also, indirectly, rehabilitating and assisting them amongst their relatives and friends instead of having them institutionalized or have them move to towns where they become beggars. Whenever possible we try to employ mothers or fathers so that the whole family benefits. We have had situations where the handicapped persons (say blind) is assisted by his/her able-bodied companion.

The nursery attendants supervise the operations at the nursery help keep records, make monthly reports and issue the trees to the members of the public. They also teach newcomers to the nursery the basics of how to produce seedlings for sale to the green belt movement. We noted that when the community identifies person who could play this role they would mostly identify a very poor parent whose children may be having problems with school fees.

Besides the very poor and the handicapped, school leavers are hired as promoters and follow-ups.

The promoters go ahead of everybody else in the field and talk to the members of the community about the problems of desertification giving suggestions on what they can individually do, encouraging them to dig holes and apply manure to them. They send monthly reports on the number of holes they check and issue approval tokens which the applicants take to the nurseries so that they can be issued with trees. The green belt movement expands as fast and as well as the promoters can effectively push it to the members of the public.

The follow-ups attend to the planted trees to ensure that they are indeed planted, they are being attended to and they are therefore surviving. They also send in monthly reports on the number of trees issued and the number surviving at the green belts.

A few fairly trained individuals are engaged as supervisors in the field. When operating at peak there may be 250-300 Individuals of these categories earning their living from this programme. If we were able to substantially create more jobs in the rural areas we would help in curbing migration into the urban areas in search of jobs. Most migrants are the youth and the rural poor. Migration into the urban centers only serves to aggravate the unemployment situation and the problem of the urban poor who live in shanties and city peripheral areas.

7. One of the most obvious results of deforestation and bush clearing is the soil erosion. During the rainy seasons rivers are red with the top soil. Lost top soil leaves behind impoverished sub-soil which cannot support agriculture and as a result food production goes down. Education is necessary so that farmers can appreciate the relationship between soil erosion and poor agricultural output.

8. Deforestation and bush clearing has precipitated an energy crisis because wood fuel has become scarce. Fetching of wood and preparation of food for the family is a responsibility of the women. And so as wood disappears women and children walk further and further from home to look for firewood which may only turn out to be twigs and sticks. Where these do not exist they will turn to agricultural residue and cow dung. These are products which should be returned to the soil in order to make it richer for food production. Burning these breaks the carbon cycle and creates a vicious cycle in agricultural production.

9. The crisis of wood fuel precipitates another problem: malnutrition. A woman with little wood fuel opts to give her family food that requires little energy to prepare. If she has money she

often turns to refined foods like bread, maize meal, tea and soft drinks. A woman may not appreciate what she must give her family to ensure a balanced diet. That ignorance, coupled with shortage of wood fuel provides an excellent background for undernourishment and diseases associated with poor feeding habits. If too many people are caught up in this situation one can easily have a sick society and a sick society in unproductive. Unproductive people are eventually pushed down into the world of underdevelopment. It is very important therefore, that the energy crisis of the poor is solved through provision of the wood and utilization of more efficient combustion devices which reduce wood consumption.

10. Indirectly, the project has been promoting a positive image of women which is a concern for the NCWK which strives to promote a balanced development of a woman's personality and to facilitate an environment in which such development can take place. Even after 10 years of debate on women issues during the women decade it appears appropriate for women to talk around development issues and cause positive change in themselves and country. Development issues provide a good forum for women to be creative, assertive and effective leaders and the green belt movement, being a development issue, provided the forum to promote women's positive image.

This is very important because women have to become involved in development as equal participants and benefactors. For currently, although women are the most numerous voters very few women are voted into public offices. This is partially because women are not afforded a forum to develop leadership qualities as they mature and even during adulthood. They are always the followers but never the leaders. Women are therefore, too often only nominated by men to positions of responsibilities. Women have always played a major role in the socio-economic and political arena of nations but they are not always publicly acclaimed, appreciated or proportionately rewarded. Indeed women are often silenced by small token positions of influence and responsibility while men are rewarded with positions they hardly deserve. Women have generally come to accept that they have to be extremely grateful for the very little they get from men both in private and public form. Those women who would point out the continued disproportionate representation of women in the decision-making structure (both political and economic) are conveniently given such labels as rebels, radicals, women libers, women elite and so forth. This is deliberately done to discredit them in the eyes of the public so that whatever they have to say or stand for is suspiciously scrutinized and preferably scorned upon. Because of this the majority of women will opt for practices which dehumanize them and make them weak, unchallenging servants to their men folk rather than partners in development. As in other areas of inequality deliberately promoted the myths of the inferiority of women can only be demolished though glaring examples with which nobody can intelligently argue. The green belt movement and other projects initiated by women are some examples around which kitchens, babies nappies and sex are not the points of reference.

D.

Have we achieved our short and long-term objectives? Most of the short-term objectives have been realized. Some of these are:

(a) To encourage tree planting so as to provide the source of energy in the rural areas.

(b) To promote planting of multipurpose trees with special reference to nutritional and energy requirements of man and his livestock.

(d) To promote the protection and maintenance of the environment and development through seminars, conferences, workshops etc.

(e) To encourage soil conservation land reclamation and rehabilitation through tree planting.

(f) To develop methods for rational land use.

(g) To create an income-generating activity for rural women.

(h) To create self-employment opportunities especially for handicapped persons and the rural poor.

(i) To develop a replicable methodology for rural development.

(j) To carry out research in conjunction with the University of Nairobi and other research institutions.

(k) To create self-employment opportunities for young persons.

(l) To carry out any activities that promote those objectives

- Thousands of trees have been produced by women planted by communities and school children in over 700 public green belts. Thousands of individuals have established private green belts. Tree planting has become an honourable activity for all and because the political leadership publicly supports conservation and reforestation efforts the general populace is easily persuaded. The sight of the President planting a tree and urging others to do the same is a valuable example for his people to emulate.

- Community tree nurseries operated by women groups, youth clubs and schools have been established in many parts of the country. At the moment they are about 50. Not only are the trees generating income for the producers but relevant knowledge is being imparted to them during demonstration sessions and visits by the trained personnel.

- Scores of individuals especially the poor and the handicapped have found jobs within their own environment amongst friends and relatives. Some children have completed their school because their parents were employed as green belt rangers.

E.

Why has this approach worked?

Many people have wanted to know why the approach we have opted for has worked. There is a combination of reasons. Some of the more obvious have been as follows:

- The green belt movement pursues several goals at the same time and focuses on several target groups all of whom can find their place in the movement.

- The short-term objectives are realized fast enough to maintain momentum and interest. People need success stories to believe.

- The Executive Committee of NCWK and those directly charged with the responsibility of guiding the movement have been very committed.

- There has been a good understanding of the issues involved. The leaders appreciated the cost of the high rate of population growth against the scarce land resource, they knew of the diminishing forest cover, they appreciated that the elimination of indigenous trees would precipitate a changed ecosystem. They felt that because they knew it was their responsibility to initiate action.

F.

Have we encountered problems? Yes indeed.

By far our greatest problem has been lack of sufficient funds to allow the programme to expand as fast as demanded by the people. The second handicap was lack of appreciation of proper planting needed and after care of trees. Perhaps our third major hurdle is the difficulty of procuring accurate records. We must be told the truth from the field. The truth may mean less money, loss of an income and sheer hard work. Working at this truth can be taxing, time consuming not to mention the fact that it can be very expensive. The need for it is not always appreciated and can develop into a major hurdle.

G.

Who has funded the Movement?

Initially, we worked with purely voluntary service which in our country is known as harambee. Then we introduced the idea of sponsoring trees which we would plant and take care of. Some substantial donations came from Mobil Oil (K) Ltd., the Environment Liaison Center, the Canadian Embassy, the German Embassy and the International Council of Women. The total amounted to Kshs. 160,000 (about US Dollars 10,000). In 1981 we hit a Jack Pot and received Kshs. 1 million (US Dollars 100,000) from the Voluntary Fund of the United Nations, ½ million (US Dollars 50,000) from the Norwegian Forestry Society and Norad and Kshs. 3 1/2 million (US Dollars 300,000) from the Danish Voluntary Fund for Developing Countries. All the grants have run currently and are scheduled to end before or in 1985. We have just received financial support from Norad of Kshs. 1.9 million (US Dollars 127,000).

H.

What we do with the funds

We purchase tools for tree nurseries and green belts, organize workshops and seminars for new participants, purchase seedlings from seedling producers (mostly women), pay green belt staff (nursery attendants, promoters, follow-ups, green belt rangers and supervisors) and maintain a small secretariat at the headquarters. The ordinary people contribute in kind by:

Digging holes for tree planting

Providing manure

Sheltering, protecting and watering the trees

Preparing nursery sites including making of benches, seedbeds etc. collecting seeds.

Many of our members supervise the operations in the fields where they are and assist with on site training for new participants.

I.

I am often asked, "Why did it take the women to start the green belt movement?" The inspiration did not come to me because I was a woman. It came to me because my mind was searching for a solution to a very specific problem. Inspirations come to all of us but many of us may not have the right mental peace and tranquility at the critical time to allow the inspiration to grow beyond the stage when it appears like a dream. I think I was just lucky. I do not know why I nursed the inspiration until it became an idea and finally an activity.

I think that women in the NCWK were quite good at pursuing that idea which for a long time bore little fruit. But that patience is not a prerogative of women. Men could have done the same if similarly inspired and sufficiently motivated. Perhaps the only thing that was characteristically women-like was our grouping and our rapid acceptance of the movement. But some observers claim that the motivating force in the field especially among women was the financial gain. May be, may be not. But if it is very few men were so motivated until much, much later.

J.

Liaison has been essential because of the nature of the project. The green belt movement has worked closely with the Ministry of Environment & Natural Resources from the very beginning. At a very early stage it was possible for our participants to walk into the forester's office and receive as many seedlings as had been prepared for. Most, if not all, foresters have co-operated in this endeavor and all appreciate the complimentary and rather unique contribution being made by the green belt movement.

There has also been very close co-operation with the office of the President (Administration) who have assisted at the district and locational levels. The green belt staff are often invited to Chief's meetings to explain the movement to the people.

Each green belt or tree nursery is supervised by a local committee comprised of leaders from the local community. This is the committee which maintains the spirits of interest and awareness after the NCWK's launching party has gone. It is the nucleus around which the community will continue to be motivated and involved. Under the leadership of the local green belt committee the community volunteers to dig holes, place manure in holes and only wait for the launching ceremony. Under the current methods such work, which would otherwise cost the tax-payer a lot of money and time is given free-of-charge by the community. K.

What of the Future?

We must continue to care and bother about issues which are not immediately concerned with the gratification of our physical senses. We are a unique heritage to the ecosystem on this planet earth and we have a special responsibility. If to those to whom more has been given more will be expected then we must embrace our special responsibility which is more than is expected of the elephants and the butterflies. In making sure that they and their future generations survive we shall be ensuring the survival of our own species. Where people have been insensitive to the life of trees, of the life that flourishes in the top soil of cropland, of life of grass and shrubs, of young children, ... yes, of all living things we have witnessed indiscriminate deforestation, soil erosions over-grazing, over-population, drought, desertification, famine and death.

More than 60% of Kenya's land is no longer available to the farmer, forests stand at low level of 25%, some river levels have fallen to minimum low before they disappear altogether. Crop yield have fallen, livestock industry is not what it used to be and our towns have many who are poor and unemployed. But we continue to cross bridges in our beautiful cars and aeroplanes and only give a passing glance to the study waters below, we cut our age-old indigenous trees to replace them with fast growing and economically valuable exotic ever-greens and we refuse to exert pressure where we should to avoid being unpopular and unsang. We are insensitive to the life of those others and we are perhaps ignorant of how much our own life depends on theirs. Yet Kenya is not among the worst in Africa. And so we must go beyond Kenya and help raise public awareness in other parts of the Continent.

The financial reward will be used to establish a Trust which could be used to provide seed money for the establishment of programmes similar to the green belt movement elsewhere in Africa. We are confident that once we start such an effort would appeal to others who are concerned about the desertification processes, prolonged drought, famine and death in our region.

I thank you therefore, on my own behalf, on behalf of all the beneficiaries both current and those in years to come. I thank you for caring, for appreciating and for rewarding.

I am encouraged, strengthened and inspired by your kindness and generosity of heart and mind.

Thank you most sincerely.

APPENDIX B

THE BOTTOM IS HEAVY TOO: EVEN WITH THE GREEN BELT MOVEMENT –

THE FIFTH EDINBURGH MEDAL ADDRESS

I am deeply honoured today. I have come here to receive an award both on my own behalf and on behalf of thousands of grassroots women, men and children with whom I have spent the last twenty years of my life in a partnership intended to utilise my education and experience to better the quality of life of my family and my community in particular, and my country and the world in general. The partners represent the people at the bottom of a pyramid which local and international political and economic systems have created on both sides of the equator. At the bottom of the pyramid are people of all shades, races, religions and gender but by far the majority at the bottom are citizens of the world south of the equator.

Upon birth, we begin a journey which should lead to happiness and fulfilment. That is the purpose of all our efforts. Between birth and death, however, there are many obstacles which separate us from that goal. Some are natural but most are man-made. The fulfilment and happiness we crave for on this planet should be possible and there should be enough for everyone's needs. Believing it to be so, we wake up every morning to toil on the resources available to us so that we can realize the goal of happiness and fulfilment. For many of us, however, and especially those at the bottom of that pyramid, there are not enough resources to meet even our basic needs.

The greater awareness we now have of the systems of our planet earth makes us appreciate the dilemma which Rachel Carson described in her book *Silent Spring* before it became fashionable to appear green. The air we breathe, the water we drink and the soil in which crops and other vegetation grow are limited resources. They are available to us to use and achieve fulfilment and happiness, as others before of us have done and those who will follow us will need to do. It was Mahatma Gandhi who gave the world the often-quoted words of wisdom, 'the world has enough for everyone's need but not for everyone's greed'. These words become even more meaningful the more we appreciate that this planet is a closed system and therefore what there is on the earth is all we've got! How then do we achieve our goals if we feel the need to satisfy not only our needs but also our greed? How can we make this journey and realise our goal despite our limited resources? After all in some countries and for the majority of people at the bottom of the pyramid the journey comes to an end even as it begins and just as most of us turn 35 or thereabouts.

The journey was often made in fear and ignorance until scientific knowledge helped us understand our world better so that we could overcome our ignorance and therefore our fear. We now discover ourselves and see where we are in relation to the rest of creation and what we ought to do as we seek fulfilment and happiness. Some of us even believe now that perhaps we are not the only ones making this journey. Perhaps all creation has a purpose and this has nothing to do with our own happiness and fulfilment. Perhaps it is not our business to decide which species should be allowed to exist and which should be denied this right, just because they are no use to us just now.

To understand our position in relation to other forms of creation, we should seek knowledge and inspiration from science and from creation itself. The message coming to more and more people now strongly suggests that we are one species which needs to be less arrogant and exploitative against what St Francis called our brothers and sisters in the wide spectrum of creation. Every other species has a right to exist and to pursue its own happiness and fulfilment and has no obligation to homo sapiens. The species should assist each other and help each other to achieve the goal of happiness and fulfilment. Homo sapiens, by virtue of its higher intelligence and a capacity for love and compassion should be more a custodian and less the exterminator.

Science and technology now dominate and have greatly changed almost all aspects of our lives on both sides of the equator. This is especially true at the bottom of the pyramid where applied modern science is a new experience and scientific and technological know-how is lacking. Indeed the bottom of the pyramid associates science and technology with magic and miracles of the glittering industrialised world. And with good reason: commercialised science has greatly enriched societies which have made scientific discoveries and have been able to apply them and create new and more efficient tools. This miracle appears well beyond reach of the bottom of the pyramid and may even be perceived as magic or a gift from God. Even though science and technology impacts on the world at the bottom of the pyramid that world hardly understands them or how the impact comes about. The people toil nonetheless, and search for their own happiness and fulfilment. But is it possible for them to realise their goal with so few resources and without science and technology? And will those who have this know-how be willing to share it when it gives them the advantage over the bottom? How can they when with that advantage they (the top) can exploit not only their own resources but also the untapped resources belonging to those at the bottom of the pyramid?

The world I work in is concerned about the environment. The resources on the planet earth are not only limited, they are also being degraded. The people at the bottom of the pyramid do not understand limits to growth and they do not appreciate that as they seek their own happiness and fulfilment they could adversely affect the same resources and jeopardise the capacity of future generations to meet their own needs. And having said that of the bottom of the pyramid, the top is blinded by insatiable appetites backed by scientific knowledge, industrial advancement, the need to acquire, accumulate and overconsume. The revolution in information dissemination worldwide plays on the ignorance and the fear of those at the bottom of the pyramid. It promotes the lifestyle of those at the top of the pyramid and sells it as the ultimate in fulfilment and happiness.

In my part of the world, environmental degradation is brought about by soil erosion, deforestation, pollution and loss of biological diversity in our earth systems. These in turn are brought about by political and economic policies and activities which are dictated by greed, corruption, incompetence and an insatiable desire to satisfy the inflated egos and ambitions of those who wield political and economic power. They are exacerbated by population pressure, international debts and interest rates, low prices for export goods, commodity protectionism and inevitable and debilitating poverty.

Many governments, aid agencies and charitable organisations invest heavily in the symptoms of environmental degradation as they mop up the world. Less effort and enthusiasm is demonstrated in dealing with the causes of the garbage they are so willing to mop up. This is not to say that people are not appreciative of aid and charity. But the majority of the people at the bottom of the pyramid are both the causes and the symptoms of environmental degradation. They are caught in a vicious cycle of poverty and underdevelopment. The Green Belt Movement endeavours to assist them to break that cycle and liberate themselves from the bonds which block their paths and separate them from their goal of happiness and fulfilment. Lifting them may be a noble and fulfilling challenge but it is also very demanding because the bottom of the pyramid, especially south of the equator, is very heavy.

Many of us at the bottom make our children believe that education is the key to a good job, a good salary and a good quality of life. They believe that education will get them out of the bottom of the pyramid and provide comfort without effort. It seems easy enough because passing examinations and moving to the next grade may come easily. As they struggle through school they console themselves with the promised success which will ensure them a place at the top of the pyramid. If that depended on good grades and certificates many of us would have little to worry about. We would be at the top!

Between reality and childhood dreams are many man-made hurdles which the people at the bottom fight against all their lives as they try to overcome them and to achieve meaningful development, improve their quality of life and realise full potential. These obstacles prevent them from utilising much of the knowledge, expertise and the experience they have acquired in their studies and in the course of their lives. This knowledge and experience is supposed to make the journey surer and easier. But there is a big difference between childhood dreams and the reality of the pyramid. At the bottom of the pyramid, sooner or later we all learn that.

Take me for example. I am basically a biologist. However, in the course of my post-graduate work I acquired experience in histological preparations of laboratory specimens and basic principles of embryology. With that background I was hired by Professor Dr R R Hofmann, who became my academic adviser and friend, to teach micro- and developmental anatomy in the faculty of Veterinary Medicine at the University of Nairobi. I felt satisfied with playing an important role in educating future veterinarians who would supervise the livestock industry in my country. Such experts were expected to ensure that there would be adequate and healthy animal production provision for our society, to control animal diseases and ensure that our livestock industry was successful.

Part of the university assignment is to do research and publish results in scientific journals so that higher academic promotions can be achieved. Eager to make my mark in the scientific world, and of course also earn my academic credentials, I commenced on a research project immediately. I decided to work on a problem which was adversely affecting the livestock industry and especially the dairy section. In order to improve our indigenous dairy cattle we had imported exotic breeds from Europe and were crossbreeding them with local stock. The project was very successful except for one problem: east coast fever. This parasitic disease proved 100% fatal to the imported exotic breeds and their progenies.

The parasite is transmitted from one animal to another by brown ear ticks, so called because the ticks love to congregate particularly on the ears of the victim. The parasite is ingested by the tick from an infected animal during feeding and eventually finds its way into the salivary glands of the ticks. From here the parasite is passed onto the next victim during the next feeding. I decided to work on the microscopic anatomy of this parasite because I was keen to make a contribution towards its life cycle. I started with its anatomy in the salivary glands of the infected ticks.

Anxious to be a good career woman and set a good example to fellow members of my gender, students and colleagues who had not worked with women professionals before, I did what I thought mattered: I reported to work on time and was both industrious and productive. Upward mobility seemed assured if the university authority would respect what they had written in the letters of my appointment! But the inevitable happened: there was hurdle which nobody articulated. It was not an academic hurdle, but a hurdle nevertheless. Mobility upwards was too slow. It was as if I did not matter as much as the others. There was something I did not have and

I could not have. The hurdle had nothing to do with passing examinations, having certificates or being a good teacher. It had everything to do with my gender! What a discovery!

I had just returned from the United States of America where I spent the first of part of the 1960s. Those years are partly remembered as the years of the Civil Rights Movement which was led by Martin Luther King Jnr. At least in the street battles the issues were clear: colour was the problem. Several years later I was in the village of my birth and childhood and I was at home with people who were black like me. I was still not o.k. This time though, it was my gender that was the problem. I have since learnt that at the bottom of the pyramid there are very strict cultural and religious norms which govern the birth, life and death of women in society. These age-old traditions make the bottom quite heavy.

By now I had tied a knot with a man. To do everything right we followed all the proper religious and traditional requirements. He promised happiness and fulfilment. He was a good Christian like me, had also been educated abroad, had been exposed to western ideas and values and we shared our traditional wisdom. I never would have thought that all the things I had worked so hard for in school and at home would become a burden, an obstacle to my domestic peace. Apparently, those academic certificates and letters of appointment to high offices were secretly emasculating the man in my life! What a catastrophe! I should have known that ambition and success were not expected to be a dominant character in, especially, an African woman. An African woman should be a good African woman whose dominant qualities should include coyness, shyness, submissiveness, incompetence (feigned if necessary) and crippling dependency if they have the opportunity to be economically independent. A highly educated, independent African woman is bound to be dominant, aggressive, uncontrollable, a bad influence on other African women. She is unmarriageable! Such qualities are attributes expected in men only. I lamented that nobody warned me about such hurdles! (A large part of those at the bottom of the pyramid still struggle to keep the lid on.) In the meantime I struggled for freedom so that I could realise my full potential, but so many opportunities to improve the bottom had been lost, much energy wasted and a lot of mileage on the journey lost. The bottom is especially very heavy for women.

To go back to my research project, in the early 1970s I spent much time collecting ticks which should have been infected with the east coast fever parasites. The cows which carried the ticks were often so skinny that I could count their ribs. This was because there was inadequate grass for them and they obviously did not have enough supplements. This observation eventually led me to appreciate the relationship between the well being of domestic animals, a degraded environment and the carrying capacity of any resource. A degraded environment could not sustain our domestic animals. Indeed the livestock industry was threatened more by environmental degradation than by either the ticks on the ears of the animals or the cast coast fever parasites in the salivary glands of ticks.

That was one of the many experiences which led me into environmental activism. I henceforth sought to understand and appreciate the symptoms as opposed to the causes of environmental degradation. This and others concerns inspired me to initiate the Green Belt Movement. The overall objective of the Movement was to raise awareness of symptoms of environmental degradation and raise the consciousness of people to a level that would move them to participate in the restoration and the healing of the environment. The majority of the people at the bottom of the pyramid would rather deal with the symptoms because their objectives are short-term and are directed towards immediate survival. The Green Belt Movement encourages them to understand the need to get to the root causes and to act.

But the women of the Green Belt Movement try. To begin with they are mostly rural women who can hardly read or write their mothers tongues, let alone the official and national languages of Kenya, namely English and Kiswahili. And there are about forty-two different mother tongues in Kenya. Communication is therefore a big barrier and although practical teaching by demonstration is applied, there is not enough time and personnel to go around. Our programme does not incorporate adult education but there are many groups which plant trees and also participate in the adult education programme conducted by the Ministry of Education. Fortunately, forestry techniques are simple and are similar to the practices applied by the farmers. With basic demonstrations groups of women are able to adapt the various forestry techniques and to overcome many problems which could be a nightmare to a professional forester.

The Green Belt Movement is basically an environmental campaign for tree planting. The objectives are many and varied but the overall concern is to raise awareness of ordinary men and women of the need to take care of the environment so that it in turn can take care of at least their basic needs. The initiators are groups of women who mainly come forward because they experience the direct impact of an environment which is degraded. They lack wood fuel, water, food and fodder. They are poor, have no cash income and are confined to rural life. They work very hard. For example, in sub-Sahara Africa they produce 80% of the food, provide the manual labour on farms and homes, raise their many children and serve as heads of households for their absent husbands. Yet they form the bulk of the bottom of the pyramid. Without education, capital, political and economic policies to support them they find themselves engulfed in vicious cycles of debilitating poverty, lost self confidence and a never-ending struggle to meet their most basic needs.

For the past fifteen years the Movement has been trying to break that cycle. The greatest obstacles have been the very systems which are created by the people at the top. These systems are designed to disempower them, to deny them basic freedoms and rights. This is done so that those at the top can more easily rule over and continue to exploit them. Because of trying to uplift the bottom of the pyramid, the Movement has been portrayed as anti-Government and the organisers and partners as dissidents. I have been the subject of unsavoury and even uncouth commentary and have been threatened with bodily harm by the political leaders who swear to protect a constitution in which are enshrined the right to freedom of movement, information, expression and association! The rights of those at the bottom of the pyramid are violated every day by those at the top.

The sheer number of those at the bottom of the pyramid creates the weight. This is compounded by all the problems enumerated above. And the numbers are getting bigger. The economic and political systems are designed to create more numbers, population pressures show no signs of waning, deforestation and desertification and other aspects of environmental degradation continue. The signs are everywhere on the wall.

Science and technology can lighten the burden but it is not being given a chance. Perhaps part of the answer lies with man itself. Humans have to reassess their roles on this planet, reassess their values, reassess their understanding of the universe and perception of what constitutes their happiness and fulfilment. We may have to reassess our system of governance and seek security and peace not in a pyramid but in a balanced and harmonious whole. For as long as we sustain a pyramid the bottom will continue to gather momentum and may take all of us with it where it is always going... to the abyss of the bottom!

APPENDIX C

THE NOBEL PEACE PRIZE LECTURE

Your Majesties

Your Royal Highnesses

Honourable Members of the Norwegian Nobel Committee

Excellencies

Ladies and Gentlemen

I stand before you and the world humbled by this recognition and uplifted by the honour of being the 2004 Nobel Peace Laureate.

As the first African woman to receive this prize, I accept it on behalf of the people of Kenya and Africa, and indeed the world. I am especially mindful of women and the girl child. I hope it will encourage them to raise their voices and take more space for leadership. I know the honour also gives a deep sense of pride to our men, both old and young. As a mother, I appreciate the inspiration this brings to the youth and urge them to use it to pursue their dreams. Although this prize comes to me, it acknowledges the work of countless individuals and groups across the globe. They work quietly and often without recognition to protect the environment, promote democracy, defend human rights and ensure equality between women and men. By so doing, they plant seeds of peace. I know they, too, are proud today. To all who feel represented by this prize I say use it to advance your mission and meet the high expectations the world will place on us.

This honour is also for my family, friends, partners and supporters throughout the world. All of them helped shape the vision and sustain our work, which was often accomplished under hostile conditions. I am also grateful to the people of Kenya - who remained stubbornly hopeful that democracy could be realized and their environment managed sustainably. Because of this support, I am here today to accept this great honour.

I am immensely privileged to join my fellow African Peace laureates, Presidents Nelson Mandela and F.W. de Klerk, Archbishop Desmond Tutu, the late Chief Albert Luthuli, the late Anwar el-Sadat and the UN Secretary General, Kofi Annan.

I know that African people everywhere are encouraged by this news. My fellow Africans, as we embrace this recognition, let us use it to intensify our commitment to our people, to reduce conflicts and poverty and thereby improve their quality of life. Let us embrace democratic governance, protect human rights and protect our environment. I am confident that we shall rise to the occasion. I have always believed that solutions to most of our problems must come from us.

In this year's prize, the Norwegian Nobel Committee has placed the critical issue of environment and its linkage to democracy and peace before the world. For their visionary action, I am profoundly grateful. Recognizing that sustainable development, democracy and peace are indivisible is an idea whose time has come. Our work over the past 30 years has always appreciated and engaged these linkages.

My inspiration partly comes from my childhood experiences and observations of Nature in rural Kenya. It has been influenced and nurtured by the formal education I was privileged to receive in Kenya, the United States and Germany. As I was growing up, I witnessed forests being cleared and replaced by commercial plantations, which destroyed local biodiversity and the capacity of the forests to conserve water.

Excellencies, ladies and gentlemen,

In 1977, when we started the Green Belt Movement, I was partly responding to needs identified by rural women, namely lack of firewood, clean drinking water, balanced diets, shelter and income.

Throughout Africa, women are the primary caretakers, holding significant responsibility for tilling the land and feeding their families. As a result, they are often the first to become aware of environmental damage as resources become scarce and incapable of sustaining their families. The women we worked with recounted that unlike in the past, they were unable to meet their basic needs. This was due to the degradation of their immediate environment as well as the introduction of commercial farming, which replaced the growing of household food crops. But international trade controlled the price of the exports from these small-scale farmers and a reasonable and just income could not be guaranteed. I came to understand that when the environment is destroyed, plundered or mismanaged, we undermine our quality of life and that of future generations.

Tree planting became a natural choice to address some of the initial basic needs identified by women. Also, tree planting is simple, attainable and guarantees quick, successful results within a reasonable amount time. This sustains interest and commitment.

So, together, we have planted over 30 million trees that provide fuel, food, shelter, and income to support their children's education and household needs. The activity also creates employment and improves soils and watersheds. Through their involvement, women gain some degree of power over their lives, especially their social and economic position and relevance in the family. This work continues.

Initially, the work was difficult because historically our people have been persuaded to believe that because they are poor, they lack not only capital, but also knowledge and skills to address their challenges. Instead they are conditioned to believe that solutions to their problems must come from 'outside'. Further, women did not realize that meeting their needs depended on their environment being healthy and well managed. They were also unaware that a degraded environment leads to a scramble for scarce resources and may culminate in poverty and even conflict. They were also unaware of the injustices of international economic arrangements.

In order to assist communities to understand these linkages, we developed a citizen education program, during which people identify their problems, the causes and possible solutions. They then make connections between their own personal actions and the problems they witness in the environment and in society. They learn that our world is confronted with a litany of woes: corruption, violence against women and children, disruption and breakdown of families, and disintegration of cultures and communities. They also identify the abuse of drugs and chemical substances, especially among young people. There are also devastating diseases that are defying cures or occurring in epidemic proportions. Of particular concern are HIV/AIDS, malaria and diseases associated with malnutrition.

On the environment front, they are exposed to many human activities that are devastating to the environment and societies. These include widespread destruction of ecosystems, especially through deforestation, climatic instability, and contamination in the soils and waters that all contribute to excruciating poverty.

In the process, the participants discover that they must be part of the solutions. They realize their hidden potential and are empowered to overcome inertia and take action. They come to recognize that they are the primary custodians and beneficiaries of the environment that sustains them.

Entire communities also come to understand that while it is necessary to hold their governments accountable, it is equally important that in their own relationships with each other, they exemplify the leadership values they wish to see in their own leaders, namely justice, integrity and trust.

Although initially the Green Belt Movement's tree planting activities did not address issues of democracy and peace, it soon became clear that responsible governance of the environment was impossible without democratic space. Therefore, the tree became a symbol for the democratic struggle in Kenya. Citizens were mobilised to challenge widespread abuses of power, corruption and environmental mismanagement. In Nairobi 's Uhuru Park, at Freedom Corner, and in many parts of the country, trees of peace were planted to demand the release of prisoners of conscience and a peaceful transition to democracy.

Through the Green Belt Movement, thousands of ordinary citizens were mobilized and empowered to take action and effect change. They learned to overcome fear and a sense of helplessness and moved to defend democratic rights.

In time, the tree also became a symbol for peace and conflict resolution, especially during ethnic conflicts in Kenya when the Green Belt Movement used peace trees to reconcile disputing communities. During the ongoing re-writing of the Kenyan constitution, similar trees of peace were planted in many parts of the country to promote a culture of peace. Using trees as a symbol of peace is in keeping with a widespread African tradition. For example, the elders of the Kikuyu carried a staff from the *thigi* tree that, when placed between two disputing sides, caused them to stop fighting and seek reconciliation. Many communities in Africa have these traditions.

Such practises are part of an extensive cultural heritage, which contributes both to the conservation of habitats and to cultures of peace. With the destruction of these cultures and the introduction of new values, local biodiversity is no longer valued or protected and as a result, it is quickly degraded and disappears. For this reason, The Green Belt Movement explores the concept of cultural biodiversity, especially with respect to indigenous seeds and medicinal plants. As we progressively understood the causes of environmental degradation, we saw the need for good governance. Indeed, the state of any county's environment is a reflection of the kind of governance in place, and without good governance there can be no peace. Many countries, which have poor governance systems, are also likely to have conflicts and poor laws protecting the environment.

In 2002, the courage, resilience, patience and commitment of members of the Green Belt Movement, other civil society organizations, and the Kenyan public culminated in the peaceful transition to a democratic government and laid the foundation for a more stable society.
Excellencies, friends, ladies and gentlemen,

It is 30 years since we started this work. Activities that devastate the environment and societies continue unabated. Today we are faced with a challenge that calls for a shift in our thinking, so that humanity stops threatening its life-support system. We are called to assist the Earth to heal her wounds and in the process heal our own – indeed, to embrace the whole creation in all its diversity, beauty and wonder. This will happen if we see the need to revive our sense of belonging to a larger family of life, with which we have shared our evolutionary process.

In the course of history, there comes a time when humanity is called to shift to a new level of consciousness, to reach a higher moral ground. A time when we have to shed our fear and give hope to each other.

That time is now.

The Norwegian Nobel Committee has challenged the world to broaden the understanding of peace: there can be no peace without equitable development; and there can be no development without sustainable management of the environment in a democratic and peaceful space. This shift is an idea whose time has come.

I call on leaders, especially from Africa, to expand democratic space and build fair and just societies that allow the creativity and energy of their citizens to flourish.

Those of us who have been privileged to receive education, skills, and experiences and even power must be role models for the next generation of leadership. In this regard, I would also like to appeal for the freedom of my fellow laureate Aung San Suu Kyi so that she can continue her work for peace and democracy for the people of Burma and the world at large.

Culture plays a central role in the political, economic and social life of communities. Indeed, culture may be the missing link in the development of Africa. Culture is dynamic and evolves over time, consciously discarding retrogressive traditions, like female genital mutilation (FGM), and embracing aspects that are good and useful.

Africans, especially, should re-discover positive aspects of their culture. In accepting them, they would give themselves a sense of belonging, identity and self-confidence.

Ladies and Gentlemen,

There is also need to galvanize civil society and grassroots movements to catalyse change. I call upon governments to recognize the role of these social movements in building a critical mass of responsible citizens, who help maintain checks and balances in society. On their part, civil society should embrace not only their rights but also their responsibilities.

Further, industry and global institutions must appreciate that ensuring economic justice, equity and ecological integrity are of greater value than profits at any cost.

The extreme global inequities and prevailing consumption patterns continue at the expense of the environment and peaceful co-existence. The choice is ours.

I would like to call on young people to commit themselves to activities that contribute toward achieving their long-term dreams. They have the energy and creativity to shape a sustainable future. To the young people I say, you are a gift to your communities and indeed the world. You are our hope and our future.

The holistic approach to development, as exemplified by the Green Belt Movement, could be embraced and replicated in more parts of Africa and beyond. It is for this reason that I have established the Wangari Maathai Foundation to ensure the continuation and expansion of these activities. Although a lot has been achieved, much remains to be done.
Excellencies, ladies and gentlemen,

As I conclude I reflect on my childhood experience when I would visit a stream next to our home to fetch water for my mother. I would drink water straight from the stream. Playing among the arrowroot leaves I tried in vain to pick up the strands of frogs' eggs, believing they were beads. But every time I put my little fingers under them they would break. Later, I saw thousands of tadpoles: black, energetic and wriggling through the clear water against the background of the brown earth. This is the world I inherited from my parents.

Today, over 50 years later, the stream has dried up, women walk long distances for water, which is not always clean, and children will never know what they have lost. The challenge is to restore the home of the tadpoles and give back to our children a world of beauty and wonder.
Thank you very much.

REFERENCES

About us: Edinburgh International Science Festival. (n. d.). Retrieved from

http://www.sciencefestival.co.uk/about-us

Bateson, G. (1972). *Steps to an ecology of mind*. New York, NY: Ballantine.

Beebe, S. A., & Beebe, S. J. (2009). *Public speaking: An audience-centered approach* (7th ed.). Boston, MA: Allyn & Bacon.

Benford, R. D., & Snow, D. A. (2000). Framing processes and social movements: An overview and assessment. *Annual Review of Sociology, 26,* 611-639.

Bethell, T. (2004/2005). Nobel Farce. *The American Spectator, 37,* 58-59.

Black, J. E. (2003). Extending the rights of personhood, voice, and life to sensate others: A homology of Right to Life and Animal Rights rhetoric. *Communication Quarterly, 51,* 312-331.

Bowers, C. A., & Appfel-Marglin, F. (2004). *Re-thinking Freire: Globalization and the environmental crisis.* New York, NY: Lawrence Erlbaum.

Bowers, J. W., Ochs, D. J., Jensen, R.J., & Schulz, D. P. (2010). *The rhetoric of agitation and control.* Long Grove, IL: Waveland Press.

Brock, B. L. (1985). Epistemology and ontology in Kenneth Burke's dramatism. *Communication Quarterly, 33,* 94-104.

Burke, K. (1954). *Permanence and change: An anatomy of purpose* (3rd ed.). Berkeley, CA: The University of California Press.

Burke, K. (1966). *Language as symbolic action: Essays on life, literature and method.* Berkeley, CA: The University of California Press.

Campbell, K. K. (2006). Cultural challenges to rhetorical criticism. *Rhetorical Review, 25*, 358-361.

Campbell, K. K. & Jamieson, K. H. (1978). Form and Genre in Rhetorical Criticism: An Introduction. In K. K. Campbell & K. H. Jamieson (Eds.) *Form and Genre: Shaping Rhetorical Action* (pp. 18-25). Falls Church, VA: Speech Communication Association.

Carleone, D., & Taylor, B. (1998). Organizational communication and cultural studies: A review essay. *Communication Theory, 8*, 337-367.

Cathcart, R. S. (1972). New approaches to the study of movements: Defining movements rhetorically. *Western Speech, 36*, 32-88.

Cathcart, R. S. (1980). Defining social movements by their rhetorical form. *Central States Speech Journal, 31*, 267-273.

Chege, M. (2008). Kenya: Back from the brink? *Journal of Democracy, 19*, 125-139.

Cho, S. (2008). Politics of critical pedagogy and new social movements. *Educational Philosophy and Theory,* 1-16.

Cloud, D. L. (2001). Laboring under the sign of the new: Cultural studies, organizational communication, and the fallacy of the new economy. *Management Communication Quarterly, 15*, 268-277.

Cooke, K. (2005, March 12). Wangari Maathai: The first African woman to win the Nobel Peace Prize and founder of Kenya's Green Belt Movement tells Kieran Cooke why her work is a matter of life and death. *Financial Times*. Retrieved October 18, 2007, from ProQuest database.

Darder, A., Baltodano, M. P., & Torres, R. D. (2009). Critical pedagogy: An introduction. In A. Darder, M. P. Baltodano & R. D. Torres (Eds.). *The critical pedagogy reader* (2nd ed.). New York, NY: Routledge.

Dove, M. R. (2008). New environmentalism. *Harvard International Review, 30,* 7.

Entman, R. M. (1993). Framing: Toward clarification of a fractured paradigm. *Journal of Communication, 43,* 51-58.

Escobar, A. (1992). Reflections on "development": Grassroots approaches and alternative politics in the world. *Futures, 24,* 411-436.

Foley, G. (1999). *Learning in social action: A contribution to understanding informal education.* London, England: Zed Books.

Freire, P. (1970). *Pedagogy of the oppressed* (30th anniversary ed.). New York, NY: Continuum. Pedagogy in process: The letters to Guinea-Bissau,

Freire, P. (1978). *Pedagogy in process: Letters to Guinea-Bissau.* New York, NY: Seabury Press.

French, M. A. (1992, June 2). The women and Mother Earth: Kenya's Wangari Maathai, linking lives to the planet. *Washington Post.* Retrieved from the LexusNexus database.

Giroux, H. A. (1985). Critical pedagogy, cultural politics and the discourse of experience. *Journal of Education, 167,* 22-43.

Glenn, C. B. (2002). Critical rhetoric and pedagogy: (Re)considering student-centered dialogue. *Radical Pedagogy, 4.1.* Retrieved from http://radicalpedagogy.icaap.org/

Goffman, E. (1974). *Frames analysis: An essay on the organization of experience.* Cambridge, MA: Harvard University Press.

The Green Belt Movement: Books by Wangari Maathai. (n.d.). Retrieved from

http://greenbeltmove ment.org

The Green Belt Movement: Wangari Maathai. (n.d.). Retrieved from

http://greenbeltmovement.org

Griffin, C. L. (2009). *Invitation to public speaking* (3rd ed.). Boston, MA: Wadsworth, Cengage

Learning.

Griffin, L. M. (1952). The rhetoric of historical movements. *Quarterly Journal of Speech, 38,*

184-188.

Gruenewald, D. A. (2003). The best of both worlds: A critical pedagogy of place. *Educational*

Researcher, 32, 3-12.

Holistic. (1989, online version November 2010). In The Oxford English Dictionary Online (2nd

ed.). Retrieved from http://www.oed.com.proxy.lib.wayne.edu/view/Entry/87727

Holst, J. D. (2002). *Social movement, civil society and radical adult education.* Westport, CT:

Bergin & Garvey.

hooks, b. (1994). *Teaching to transgress: Education as the practice of freedom.* New York, NY:

Routledge.

Huesca, R. (2001). Conceptual contribution of new social movements to development

communication research. *Communication Theory, 11,* 415-433.

Hunt, S. B. (2003). An essay on publishing standards for rhetorical criticism. *Communication*

Studies, 54, 378-384.

Irons, N. (1994). Introduction to address. *The bottom is heavy too: Even with the Green Belt*

Movement – The fifth Edinburgh Medal Address. Edinburgh, Great Britain: Alexander

Ritchie & Son Ltd.

Jensen, R. J. (2006). Analyzing social movement rhetoric. *Rhetorical Review, 25,* 372-375.

Ketola, T. (2008). A holistic corporate responsibility model: Integrating values, discourses and

actions. *Journal of Business Ethics, 80,* 419-435.

Kuypers, J. A. (2005). Framing analysis. In J. A. Kuypers (Ed.). *The art of rhetorical criticism*

(pp. 186-212). Boston, MA: Pearson.

Kuypers, J. A. (2006). *Bush's war: Media bias and justification of war in a terrorist age.*

Lanham, MD: Rowman & Littlefield.

Kuypers, J. A. (2009). Framing analysis from a rhetorical perspective. In P. D'Angelo & J. A.

Kuypers (Eds.). *Doing news framing analysis: Empirical and theoretical perspectives*

(pp. 286-312). New York, NY: Routledge.

Little, J. B. (2006). Planting trees for peace. *American Forests, 112,* 47-48.

Lopez, D. S. (2001). *The story of Buddhism: A concise guide to its history and teachings.* San

Francisco, CA: HarperSanFrancisco.

Lundestad, G. (2001). *The Nobel Peace Prize, 1901-2000.* Retrieved from http://nobelprize.org/

nobel_prizes/peace/articles/lundestad-review/index.html

Maathai, W. (1984). *Right Livelihood Award Acceptance Speech.* Retrieved from http://www.

rightlivelihood.org/maathai_speech.html

Maathai, W. (1985). *The Green Belt Movement.* Nairobi, Kenya: General Printers.

Maathai, W. (1994). *The bottom is heavy too: Even with the Green Belt Movement – The fifth*

Edinburgh Medal Address. Edinburgh, Great Britain: Alexander Ritchie & Son Ltd.

Maathai, W. (2004a). *The Green Belt Movement: Sharing the approach and the experience* (2nd

ed.). New York, NY: Lantern Books.

Maathai, W. (2004b). *Nobel Lecture.* Retrieved from http://greenbeltmovement.org

Maathai, W. (2007). *Unbowed: A memoir.* New York, NY: Anchor Books.

Maliti, T. (2004). Kenyan wins Nobel Peace Prize: With a record number of nominations, the Nobel Committee opts to stay away from the politically charged Iraq war debate. *International Journal of Humanities and Peace, 20,* 9.

McLaren, P. (1999). *Schooling as a ritual performance: Toward a political economy of educational symbols and gestures* (3rd ed.). Oxford, England: Rowman & Littlefield.

McLaren, P. (2003). Critical pedagogy: A look at the major concepts. In A. Darder, M. P. Baltodano & R. D. Torres (Eds.). *The critical pedagogy reader* (1st ed.). New York, NY: Routledge.

McLaren, P. & Houston, D. (2004). Revolutionary ecologies: Ecosocialism and critical pedagogy. *Educational Studies 36,* 27-45.

Merton, L. (Producer & Director) & Dater, A. (Producer & Director). (2008). *Taking root: The vision of Wangari Maathai.* [Documentary]. United States: Marlboro Productions.

Meyer, D. S. (2007). *The politics of protest: Social movements in America.* New York, NY: Oxford University Press.

Nasong'o, S. W. (2007). Political transition without transformation: The dialectic of liberalization. *African Studies Review, 50,* 83-107.

Ndegwa, S. N. (1993). *Civil society and the promise of political development: The political impact of indigenous non-governmental organizations in Kenya* (M.A. book). Retrieved from ProQuest Dissertation database.

The Nobel Peace Prize. (2011). Retrieved from http://nobelprize.org/nobel_prizes/peace

The Nobel Prizes. (2011). Retrieved from http://nobelprize.org/nobel_prizes

Nzomo, M. (1989). The impact of the women's decade on policies, programs and empowerment of women in Kenya. *Issue: A Journal of Opinion, 17,* 9-17.

Obi, C. I. (2005). Environmental movements in sub-Saharan Africa: A political ecology of power and conflict. *UNRISD Programme on Civil Society and Social Movements, 15.*

Ott, B. L. & Aoki, E. (2002). The politics of negotiating public tragedy: Media framing of the Matthew Shepard murder. *Rhetoric & Public Affairs, 5,* 483-505.

Pineau, E. L. (2002). Critical performative pedagogy: Fleshing out the politics of libratory education. In N. Stuckey & C. Wimmer (Eds.). *Teaching performance studies.* Carbondale, IL: Southern Illinois UP.

Pinkerton, C. (2009). *Calibrating social movement theory: The politics of loving-kindness amidst the exigencies in Burma.* Paper presented at the annual conference of the International Communication Annual Conference Paper, Chicago, IL.

Plon, U. (2005, October 10). Got his eyes on the prize: Through his Right Livelihood Awards, Jakob von Uexkull honors alternative thinkers. *Time International (Europe Edition),* p. 92.

Press, R. M. (2009). Candles in the wind: Resisting in Liberia (1979-2003). *Africa Today, 55,* 2-22.

Press release – Nobel Peace Prize 2004. (2004, October 8). Retrieved from http://nobelprize.org/peace/laureates/2004/press.html

Ramphal, S. (1994). Oration. *The bottom is heavy too: Even with the Green Belt Movement – The fifth Edinburgh Medal Address.* Edinburgh, Great Britain: Alexander Ritchie & Son Ltd.

Right Livelihood Award: History – Setting up the "Alternative Nobel Prize." (n.d.) Retrieved from http://www.rightlivelihood.org/history.htm

Scatamburlo-D'Annibale, J., Suoranta, N. J., & McLaren P. (2006). Farewell to the "Bewildered Herd": Paulo Freire's revolutionary dialogical communication in the age of corporate globalization. *The Journal for Critical Education Policy Studies, 4.2.*

Sillars, M. O., & Gronbeck, B. E. (2001). *Communication criticism: Social codes, cultural studies.* Prospect Heights, IL: Waveland.

Simons, H. W. (1970). Requirements, problems, and strategies: a theory of persuasion for social movements. *The Quarterly Journal of Speech, 56,* 1-11.

Simons, H. W. (1991). On the rhetoric of social movements, historical movements, and "top-down" movements: A commentary. *Communication Studies, 42,* 94-101.

Snow, D. A., Rochford, E. B., Jr., Worden, S. K., & Benford, R. D. (1986). Frame alignment processes, micromobilization, and movement participation. *American Sociological Review, 51,* 464-481.

Stewart, C. J. (1980). A functional approach to the rhetoric of social movements. *Central States Speech Journal, 31,* 298-305.

Stewart, C. J., Smith, C. A., & Denton, R. E., Jr. (2007). *Persuasion and social movements* (5th ed.). Long Grove, IL: Waveland.

Stojanovic, D., & Gec, J. (2011, February 22). Serbian ousters of Milosevic make mark in Egypt. *The Olympian.* Retrieved from http://www.theolympian.com/2011/02/22/1552289/serbian-ousters-of-milosevic-make.html

United Nations Environment Programme (UNEP). (n.d.). The Billion Tree Campaign. Retrieved from http://www.unep.org/billiontreecampaign/

Tilly, C. (2007). Interview as cited in Press, R. M. (2009). Candles in the wind: Resisting in Liberia (1979-2003). *Africa Today, 55,* 2-22.

Toyosaki, S. (2007). Communication sensei's storytelling: Projecting identity into critical pedagogy. *Cultural Studies <--> Critical Methodologies, 7,* 48-73.

Udvardy, M. L. (1998). Theorizing past and present women's organizations in Kenya. *World Development, 26,* 1749-1761.

West, I. (2007). Performing resistance in/from the kitchen: The practice of maternal pacifist politics and La WISP's cookbooks. *Women's Studies in Communication, 30,* 358-383.

Whalen, S., & Hauser, G. (1995). Identity arguments in new social movement rhetoric. *Conference Proceedings – National Communication Association/American Forensic Association (Alta Conference on Argumentation)* (pp. 439-439). Washington, DC: National Communication Association.

Women and gender: Minister becomes continent's first woman Nobel Prize winner. (2004, October 8). *Africa News.* Retrieved October 18, 2007 from Lexis-Nexis Academic database.

Worthington, N. (2003). Shifting identities in the Kenyan press: Representations of Wangari Maathai's. *Women's Studies in Communication, 26,* 143-164.

Zarefsky, D. (1980). A skeptical view of movement studies. *Central States Speech Journal, 31,* 245-254.

Zarefsky, D. (2006) Interdisciplinary perspectives on rhetorical criticism: Reflections on rhetorical criticism. *Rhetorical Review, 25,* 383-387.